Fundamentals of Construction Safety

Fundamentals of Construction Safety

P.T. Armstrong LIOB, FTC, CGLI

Lecturer in Construction and Safety at Peterborough Technical College

Hutchinson of London

Hutchinson & Co. (Publishers) Ltd
3 Fitzroy Square, London W1P 6JD

London Melbourne Sydney Auckland
Wellington Johannesburg and agencies
throughout the world

First published 1980

Set in IBM Press Roman by Tek-Art Ltd, SE 20

Printed in Great Britain by The Anchor Press Ltd
and bound by Wm Brendon & Sons Ltd,
both of Tiptree, Essex

British Library Cataloguing in Publication data
Armstrong, P.T.
Fundamentals of construction safety.
1. Building – Great Britain – Safety measures
I. Title
624 TH443

ISBN 0 09 138410 9 cased
0 09 138411 7 paper

Contents

Acknowledgments

In addition to those mentioned personally in the Preface, the author would like to express gratitude to the following organizations for allowing him to reproduce documents:

J.W. Barber & Son Ltd, Stanground, Peterborough, for company reports and safety policy; *Bettles Building Company*, Alma Road, Peterborough, for details of safety procedures, documentation of safety department and company safety policy; *British Standards Institution,* 2 Park Street, London W1A 2BS, for extracts from British Standards (complete copies can be obtained from the BSI); *Health and Safety Executive,* George Street West, Luton, for advice and copies of documentations; *Her Majesty's Stationery Office,* Holborn Viaduct, London EC1P 1BN, for copies of Statutory Instruments and Statutory Documents; *Pneu PAC,* London Road, Dunstable, for details and photographs of automatic resuscitation; *Peterborough Technical College,* Park Crescent, Peterborough, for examples of work activities; *Stanley Powertools Ltd,* Cramlington, Northumberland, for details and information; *Stanley Tools Ltd,* Woodside, Sheffield, for information and photographs.

Preface

On the completion of this manuscript, a life's ambition has been achieved, and it is to be hoped that safety will receive greater attention throughout the construction industry as a result.

To deal adequately with this often neglected subject has been a most interesting and rewarding challenge. Readers will be varied, many of them vulnerable to dangers at work and reluctant to adapt to the changes that the law now requires.

I have enjoyed the involvement of writing this book and hope that those who read it may benefit from a safer working life.

Much help has been received from personal friends and colleagues, both in around Peterborough Technical College, and I am sincerely grateful for this. Particular thanks and appreciation are extended to Mrs Cynthia Roe for her typing and patience in deciphering my draft, and to Mr Peter Lees for his careful check-reading and subsequent advisory comments about the manuscript.

I am also truly grateful to my wife Rosalene for her constant inspiration, motivation and help in achieving this work, and I appreciate the tolerance and forbearance that she and our children, Russell and Darren, have shown throughout its fluctuating progress.

Finally, I acknowledge the support of the publishers. I trust that we shall score a mutual success and help to make the construction industry safer for both present and future employers and employees.

Illustrations

Bibliography of relevant British Standards

BS 138: 1948	Portable fire extinguishers of the water type (soda acid)
BS 196: 1961	Protected type non-reversible plugs, socket-outlets, etc.
BS 302: 1968	Wire ropes for cranes, excavators and general purposes
BS 341: 1962	Valve fittings for compressed gas cylinders
BS 349: 1973	Identification of contents of industrial gas containers
BS 638: 1966	Arc welding plant, equipment and accessories
BS 646: 1958	Cartridge fuse-links
BS 740: 1948	Part 1: Portable fire extinguishers of the foam type (chemical)
BS 740: 1952	Part 2: Portable fire extinguishers of the foam type (gas pressure)
BS 953: 1973	Methods of test for safety and protective footwear
BS 1129: 1966	Timber ladders, steps, trestles and lightweight stagings for industrial use
BS 1139: 1964	Metal scaffolding
BS 1382: 1948	Portable fire extinguishers of the water type (gas pressure)
BS 1397: 1967	Industrial safety belts and harnesses
BS 1547: 1959	Flameproof industrial clothing (materials and design)
BS 1651: 1966	Industrial gloves
BS 1721: 1968	Portable fire extinguishers of the halogenated hydrocarbon type
BS 1870: 1970	Part 1: Safety footwear other than all-rubber types
BS 1870: 1976	Part 2: Lined rubber safety boots
BS 2037: 1964	Aluminium ladders, steps and trestles for the building and civil engineering industries.
BS 2091: 1969	Respirators for protection against harmful dust, gases, etc.
BS 2092: 1967	Industrial eye-protectors
BS 2482: 1970	Timber scaffold boards
BS 2653: 1955	Protective clothing for welders
BS 2769: 1964	Portable electric motor-operated tools
BS 3016: 1972	Pressure regulators and automatic changeover devices for liquefied petroleum gases
BS 3326: 1960	Portable carbon dioxide fire extinguishers
BS 3465: 1962	Dry powder portable fire extinguishers
BS 3589: 1963	Glossary of general building terms
BS 3709: 1964	Portable fire extinguishers of the water type (stored pressure)
BS 3879: 1969	Portable liquefied petroleum gas (LPG) appliances
BS 3997: 1966	Classification of woodworking machines and auxiliary equipment
BS 4074: 1966	Metal props and struts
BS 4078: 1966	Cartridge-operated fixing tools

BS 4163: 1975	Recommendations for health and safety in workshops of schools and colleges
BS 4170: 1967	Waterproof protective clothing
BS 4171: 1967	Donkey jackets
BS 4211: 1967	Steel ladders for permanent access
BS 4310: 1968	Permissible limit of lead in low-lead points and similar materials
BS 4340: 1968	Glossary of formwork terms
BS 4343: 1968	Industrial plugs, socket-outlets and couplers
BS 4363: 1968	Distribution units for electricity supplies for construction and building sites
BS 4679: 1971	Protective suits for construction workers and others
BS 5228; 1975	Code of practice for noise control on construction and demolition sites
BS 5240: 1975	General purpose industrial safety helmets
BS 5304: 1975	Code of practice for safeguarding of machinery
BS 5378: 1976	Specification for safety colours and safety signs
BS 5426: 1976	Specification for work wear
CP 97: 1967	Part 1: Metal scaffolding: common scaffolds in steel
CP 97: 1970	Part 2: Suspended scaffolds
CP 97: 1972	Part 3: Special scaffold structures in steel

Introduction

The aim of this book is to enlighten all personnel involved in the construction industry that safety is a part of their life, and possibly death. The emphasis is intended to be direct and fundamental and the information given in a concise manner, prepared in a sequence for easy reference.

The book covers all requirements of Advanced and Basic Craft courses in respect of safety and supplements the general course structure for each trade. In addition, those in supervisory and lower management training or employment can benefit from the information, which includes all the safety requirements appertaining to employees at work and their supervisors.

All dimensions and sizes given are metric, although imperial measures are used in some current legislation. It is expected that legislation prepared and implemented prior to metrication of the construction industry will, during future updating, be amended accordingly.

All documentation included is from the sources shown, with acknowledgement as necessary. Certain areas of work are illustrated by references to 'R & D Building Company', a fictitious company.

1 Safety past and present

Had the past safety record of the construction industry not been so disastrously bad, or had it shown a glimmer of improvement, the vast legal changes concerning safety would never have been implemented. The numbers of deaths and injuries during modern construction activities are inexcusably high. During 1973 over 1000 people were killed in all industries: of these, 230 were killed in construction employment. It is not the intention of this book to warn off any present or future construction employee, nor to present frightening facts or possibilities. The facts are here to stay and can only be a grim reminder of the bad past.

Facts and figures

These figures for deaths and injuries on the British mainland are based upon Health and Safety Executive reports:

1973 231 construction workers killed and over 37 000 seriously injured
1974 166 construction workers killed and over 34 000 seriously injured
1975 182 construction workers killed and over 35 000 seriously injured
1976 156 construction workers killed and over 36 000 seriously injured
1977 130 construction workers killed and over 32 000 seriously injured

These figures make construction one of the most dangerous industrial working-places. Both management and labour in construction appear oblivious to this fact and reluctant to change their dangerous ways. Put differently the figures are:

200	killed per year
30 000	seriously injured per year
3.8	killed per week
577	seriously injured per week
0.095	killed per hour
14.4	seriously injured per hour

A brief history

A most important and powerful piece of legislation became law on 1 April 1975, when the Health and Safety at Work etc. Act was fully implemented. The following is a brief review of the law before this statutory instrument, and how it became established.

The earliest indication of legal control came during 1932 when the initial Factories Act provided for the improvement of conditions for employees, especially in the textile industries. This followed almost a century of increased industrial activity created by the drastic changes of the industrial revolution during the eighteenth century.

During the nineteenth century numerous industries came under greater legal control to eliminate unsafe practices. Construction was one of several industries that became safer and healthier as a result.

The twentieth century commenced on similar lines, with still more subsidiary pieces of legislation: legal controls were now spread among several government departments. During the latter part of the 1940s there was a change in the outlook of workers. Past were the depression years of the 1930s and the changes caused by the Second World War. The end of the war had brought about changes in three major fields: technological – bringing improved work output and modern work processes for increased production; medical – improved knowledge and understanding induced a keenness to avoid hazardous health conditions; and social – more settled conditions inspired people to be more cautious at work, to achieve a longer, happier life. These factors made for much improved worker co-operation and an increased demand for better, healthier and safer working conditions.

In the late 1960s Barbara Castle, then Minister of Labour, established a committee to search into all aspects of health and safety at work. Lord Robens was appointed Chairman of this Committee of Inquiry and the findings were published in 1972 in the Robens Report. This document was the basis of the Health and Safety at Work etc. Act, which followed in 1974.

The Health and Safety at Work etc. Act 1974

Based upon the strategy of the Robens Report, this legal document created two major conceptions to govern safe practices. Firstly, the introduction of one complete overriding law to cover comprehensively *all* persons at their place of work. This was superimposed over existing laws and was to supersede them as amendments and adjustments became necessary. Secondly, a new controlling organization unbiassed politically, was to take responsibility, under the government's Secretary of State for Employment. As a result of these recommendations there have been, to date, numerous adjustments and updatings of legislation, and this is expected to continue for several years to come. The second recommendation brought about the establishment of the following organizations, which are shown graphically in Figure 4 (page 33).

The Health and Safety Commission

All details and arrangements required by the Act are carried out by this agency, which is responsible to the relevant government ministers. Its general duties include assistance to those requesting it, along with research, training, publications and advisory services. The preparation and instigation of documents are also dealt with by the Commission in conjunction with the relevant ministers.

The Health and Safety Executive

Working upon all elements delegated by the Commission, this agency enforces the legal provisions laid down. Within its area of responsibility are the powers of Inspectors, who can issue Improvement Notices and Prohibition Notices, among other things. The Executive therefore enforces the requirements of the Commission through a procedure of mutual agreement and advantage.

The changes created by the Health and Safety at Work etc. Act

This new Act created certain new areas of control, and highlighted others that had been 'lost' within past statutory instruments. Some of the relevant factors of change or improved enforcement are:

1. All current legislation remains, and will be updated.
2. The Health and Safety Commission began work on 1 October 1974.
3. The Health and Safety Executive began work on 1 January 1975.
4. The Commission has three members from the Confederation of British Industry (CBI), three from the Trade Union Congress (TUC), two from local authorities and one from a safety organization. These nine nominees are controlled by a separate, independent chairman.
5. The Act created a new legal obligation upon all working persons. This effectively gave new legal cover for five million people.
6. Liabilities in connection with work activities have been extended, so that legal cover and obligation now include: employer, employee, suppliers/manufacturers, landlords, and the self-employed.
7. There is improved coverage of persons through the use of the phrase 'working contract' instead of 'a place of work'. Also, third parties – i.e. those affected though not directly involved – are legally protected; for example, residents living near a pollution hazard. The new wording 'health of employees', gives more coverage than the previous term, 'accident prevention'.
8. The law is now a 'statute law' (controlled by legal process) as opposed to the former 'common law'. This requires and imposes the vital clause now legally used: *'to take pre-*

cautions so far as reasonably practicable'.

9 Increased penalties for those disregarding the regulations to promote awareness and compliance.

10 Improved consultation has been developed – through prepared safety policies; better instructional literature from manufacturers; employer advising employee; inspector speaking directly to employee; development of employees' safety committees, and establishment of union representation regarding matters of safety.

The new law is now in full operation. The main function of the following chapters is to enlighten and advise the reader about how, when and by whom this law should be obeyed. This will enable lives and bodies to be saved from minor, serious or fatal injuries.

2 The employer's responsibilities

The intention of the Health and Safety at Work etc. Act 1974 was to control the activities of all employers. It would effectively protect the health, safety and welfare of persons at work through a system of compulsory control over the employer. Under this new legislation a great many employees became legally protected for the first time, workers in education, medicine, research and other fields.

Certain employments are still not covered, for example domestic employees whose place of work is a private home. There are also complicated restrictions upon those government employees considered to be 'Crown' employees. Workers in, for instance, the civil service are employed by the government (formally referred to as the 'Crown'), yet that same government is responsible for enforcing the Act upon employers – in this case upon itself. Certain restrictions and exemptions apply to Crown employees, but in general the Act applies to them as it does to others.

General duties

Each employer has now a legal obligation to maintain, so far as is reasonably practicable, the health, safety and welfare at work of his employees.

In a similar way, although not controlled by law, there exists a moral obligation for the employer to protect his employees against danger, injury or death.

The interpretation of 'reasonably practicable precautions' is: 'Take the normal, sensible actions to prevent a hazard or danger.' This wording has been chosen in preference to 'Take all possible precautions' because it cannot be possible to eliminate *all* hazards. Consider a scaffold erected and accepted by a competent person responsible, who then filled in the relevant section of Form 91 (Scaffold Register). The scaffold retained its good condition and has been inspected at the statutory seven-day intervals – all practicable precautions have been taken to ensure that it is safe. Yet the scaffold may fail. Two possibilities are:

1 Metal fatigue which visual checking cannot detect in a long-term scaffold.
2 Ground conditions beneath the surface may not be sound although thorough visual checking had been taken but failure may happen.

A *possibility* of danger always exists despite all care and preventive actions. Taking all practicable precautions eliminates all normal foreseeable hazards, so to use the scaffold following such normal practicable precautions is legally acceptable.

Detailed duties

The following are some of the detailed safety duties to be carried out by the employer to maintain acceptable protection of all his employees at work. By complying with these the employer ensures, so far as reasonably practicable, the safety of his employees. Each section is concluded with a checklist the employer may consider useful.

Plant and systems

All apparatus, machinery, equipment and plant must, at the expense and organization of the employer, be maintained in good working order that will eliminate risk to the employee. Individual construction plant and systems are detailed in later chapters; the principle is that anything used by the employee at his place of work must be maintained satisfactorily by the employer. The following checks should be considered by the employer:

1 All machines and apparatus must comply with relevant legislation, e.g. the Factories Act 1961; Woodworking Machines Regulations 1974.
2 An organized system of inspection and testing of all machines in operation by a competent person must be set up.
3 A record of all inspections by the competent person must be signed to confirm that these machine and apparatus inspections were carried out.
4 In some cases, specialists are needed to inspect equipment.
5 The work procedures must provide adequately for safety: are all safety regulations being enforced?
6 Maintenance arrangements must be organized in order to retain safety standards.
7 A good emergency procedure must exist in case of spillage, toxic gases, fire, or accident.
8 The necessary protective garments must be available at every machine with advisory notices to encourage their use. All personal issues used in conjunction with protective garments are legally required and should be implemented.
9 All operatives subject to hazard, either physical or to their health, as a result of work conditions must be advised of the problems, the correct working method and any other relevant information.
10 All precautions, so far as reasonably practicable, must be taken to maintain the health and welfare of all employees.

Storage and transport

Movement, transport and storage of materials, equipment or substances must be completed in ways which avoid risks to employees. All lifting, handling, etc., must be carried out in a safety-conscious manner, with regard to the actual work being done and also to the future movement of materials from the storage position e.g.:

1 Bricks must be stacked so that they can be removed without collapse when some are needed.
2 Unloaded timber must be stored in such a way that it can be easily taken as required.
3 Proper storage must be prepared for liquefied petroleum gas (LPG) to avoid careless use, fire risk or explosion.

Several dangers occur in each and every trade in what appear to be simple, everyday situations. It is, however, exactly those *familiar* activities that breed contempt and may produce careless attitudes, possibly resulting in injury or death. Consider the following checklist:

1 All transport and storage arrangements must be improved to create safer, healthier, conditions where necessary.
2 Employees must be adequately trained in the use of their equipment. If they are not, training must be arranged.
3 Any procedures of storage or transport that are hazardous must be improved without delay.
4 All dangerous equipment (e.g. cartridge guns) must be carefully stored, and distributed for use with caution.
5 All changes needed in the plant or buildings to maintain safety standards must be implemented without delay.
6 All containers must be suitably labelled with the contents and also with advice notes in case of spillage if they contain hazardous materials.
7 All containers and handling equipment must comply with legislation: any specialist advice must be sought if necessary to retain safe procedures.
8 Employees must be made aware of both old and new products that may be hazardous.
9 All safety garments must be available and used.
10 Accurate records of all hazardous substances must be kept as required by current legislation.

Workplace maintenance

The place at which the employee needs to work must be established and maintained by the employer. Although this is a responsibility of the employer, there is also a moral obligation upon the employee to respect and protect the workplace established for him. The term 'workplace' includes also the access to the place of work. This covers numerous situations, the most relevant to construction being on and around the building site, for example passenger hoists and the notorious hazard of moving from ladders to scaffolds. The partly erected buildings, the excavation or scaffold are the place of work: the method of physically getting there on the site is the means of access. The following checklist is typical of what an employer needs to consider.

1 Hygiene, cleanliness and general health conditions must meet legal standards.
2 Where buildings need to comply with special safety standards, specialist advice must be followed and all standards maintained.
3 There must be means of access and egress, adequate for all personnel.
4 Everyone should be aware of any especially awkward or hazardous areas of the workplace.
5 Fire exits, clearly marked and free from obstruction, must be regularly checked.
6 If the fire-fighting equipment does not meet the minimum requirements as regards suitable types of extinguisher and location of fire points, consult the local fire officer.
7 A current fire certificate for the premises, covering periodical testing of equipment, must be held or arrangements must be made to obtain one.
8 Regular fire drills must be called and observed.
9 Adequate emergency stop controls for machine areas and suitable stand-by light for power failures must be provided and regularly checked.
10 A balanced security system must be established which enables entry in an emergency but retains security.

Work conditions and environment

As well as being physically safe, the employee must not be subjected to any environmental health hazard or pollution.

Confined places such as workshops may produce health hazards from dust, toxic fumes, noise, skin diseases or ill effects from work carried out by powered hand-tools. These dangers to the employees must be eliminated or controlled to acceptable standards and protective clothing must be provided, as detailed in later chapters. The following is a typical checklist:

1 Lighting, heating and ventilation standards must conform with legal requirements.
2 Suitable, sufficient and clean sanitary facilities must be provided.
3 Washing facilities must meet the requirements of the number of people who use them.
4 There must be safe disposal of all wastes.
5 Adequate first aid facilities must be fully maintained.

Training and supervision

To achieve all this, a good standard of safety training must be offered to younger construction workers. This is provided at colleges of further education in day-release, block-release, full-time or sandwich courses. Safety is an integral part of such studies, which incorporate detailed knowledge of each trade and its interrelation and co-ordination with other trades. The training of mature and older construction workers is achieved by short courses and one-day seminars to highlight particular hazards and current safety trends. The Construction Industry Training Board has provided several courses of interest and advantage both to employer and employee. These courses, taken to the company office or site, promote safety and welfare. Both the CITB and the National Federation of Building Trades Employers run schemes with travelling lecturers and films, which bring safety knowledge to the employee instead of taking him away on a course. The legal aspects of safety are changing as progress is made towards improving standards of employee protection. Since it is part of the employer's obligation to train operatives and personnel, courses and lectures are becoming an important part of construction industry training.

In addition to this training, it is obligatory for the employer to keep up a good standard of supervision.

It is unwise, and more important unsafe, to expose a semi-skilled, unskilled or partially trained operative to a situation that may be hazardous. All operatives, however skilled or experienced need the support, guidance and control of a supervisor in most everyday situations. Supervision must be available as and when needed to maintain acceptable safety standards. An employer needs to remember the following:

1 Adequate and correct information must be provided to all levels of employee covering all legal requirements. This information must be periodically updated.
2 There must be a suitable distribution scheme to pass around all statutory information.
3 Supervisory grades of employee need to be fully trained to advise subordinates with regard to safety.
4 A scheme whereby all employees are trained to the minimum safety standards must be implemented.
5 The employer must provide all possible aids to training in the way of courses, colleges, lectures, films, etc., and see them used to the best advantage.
6 The employer must see that all employees are kept in a safety-conscious frame of mind by suitable training and guidance.
7 Where an employee is exposed to a health hazard he must be advised of the danger and how steps must be taken to eliminate his risk.
8 Trainees, fully supervised for their particular task, should receive accurate advice – with monitoring of the training and literature provided.

Safety policy

Every employer, except one employing less than five people, must prepare, and update as necessary, a written 'safety policy'. This is a detailed document outlining the intentions of the employer to implement all safety obligations and giving details of the responsibilities of both employees and employer.

Typical practical rules in such a document might be:

1 All persons working on site will wear protective helmets.
2 All operatives will be issued with goggles to the prescribed standard suitable for their activities.
3 No person will enter any excavation until given clear instruction by the competent person responsible for checking such excavation at the commencement of each working shift.
4 All persons working in the workshop, compound or yard will wear suitable protective footwear.

A typical example of safety policy is shown as Figure 11 (page 47) where the company outlines its responsibilities. Similar practical policies are implemented at site level by the site leader. All employees should be issued with a copy of the policy to ensure their full knowledge and understanding of its meaning. The following should be implemented:

1 Where the need exists, a suitable policy of safety should be prepared, approved and issued to all employees.
2 The safety policy when implemented must be maintained and observed by all employees.
3 The employees should be properly consulted regarding the safety policy.
4 All measures must be taken to encourage the safety, health and welfare of all parties.

Safety representation within the workforce

Employers are required to establish and maintain a good liaison with their workforce by the election or selection of special representatives from the workforce. This allows direct communication of requirements from the workforce to management and vice versa. Regular meetings should establish a working relationship to keep all safety measures fully known and effective. Large organizations may create larger groups or works committees which fulfil several company needs and include safety as an important part of their discussions. The following points should be remembered:

1 Adequate consultations with the employees to establish a good liaison are needed to comply with the Health and Safety at Work etc. Act.

2 Representatives, elected or selected, should meet the employer as necessary to discuss and review safety.
3 These representatives need to be fully aware of the employer's position and problems to enable a fair assessment to be made.
4 Has provision been made in the safety policy for fair representation and also for an equal contribution from all parties?

Safety representation from the unions

Safety representatives, either elected or selected from the workforce, are involved in consultation with the employer in maintaining the safety standard requirements of the Health and Safety at Work etc. Act.

Whilst it is a legal requirement that he must accept these representatives from the workforce, the employer is also expected to work in co-operation with such committees to maintain safety standards. The following should be considered:

1 There must be full and complete consultation with the Union to elect or select suitable representation.
2 A full liaison once established must be maintained with all Unions whether they have a representative on the working committee or not.
3 Is the working committee free from restrictions and allowed to work properly? Restrictions should be lifted if this is not so.

Responsibilities to self-employed

All employers have a legal obligation to ensure safe working conditions for people not employed by him who may be at risk as a result of the employer's working activities. There also exists a responsibility for self-employed persons (page 25). This checklist is typical of that required by the employer:

1 The facilities offered to those not directly employed by the employer must be of equal standard and quality to facilities offered to all other employees.
2 The self-employed persons must be aware of the safety policy, its intentions and the need for their co-operation.
3 All details of statutory instruments and safety documents must be made available to the self-employed.
4 A suitable monitoring check upon the self-employed must be maintained.
5 All activities to any person must be unbiased and clear to all parties working, regardless of employer.

3 The employee's responsibilities

It may appear from the previous chapter that the responsibilities placed upon the employee may be small compared with those of the employer. This is far from correct. In fact the employee has a vital role in maintaining the safe preservation of his health, welfare and life as well as those affected by his work activities.

The Health and Safety at Work etc. Act protects the employee against his employer's error. It also protects the employer by controlling the bad practices and errors of employees.

Consider the following definitions:

An *employer* is a person, company or organization who retains an individual in employment on a condition of a contract, and who pays that person monetary reward for work done within such terms of the contract of employment.

An *employee* is an individual who offers his skill, knowledge, experience and expertise to the employer for a return of monetary payment and an agreed contract of employment.

General duties of employee

The employee can only be responsible to his employer at the times and within the conditions of his agreed period of work. The legislation makes a strong charge to the employee to take responsibilities such as have never been equalled by any previous statutory control. There has always existed a common law obligation upon the employer to protect his employee, but this has now been intensified to an established statute law. This, in effect, means a legal protection exists for the employee whilst at work. It makes great efforts to help the employee and demands, by the same statute law, a simple and direct return of 'reasonable care'. The requirements of the employee are quite straightforward according to the Act and are:

It shall be the duty of every employee whilst at work –

a) 'to take reasonable care for the health and safety of himself and other persons who may be affected by his acts or omissions at work.

b) 'as regards any duty or requirement imposed upon his employer or by any other person or under any of the relevant statutory provisions, to co-operate with him so far as necessary to enable that duty or requirement to be performed or complied with'.

The Act therefore calls for co-operation and compels the employee to do all within his power to help. The Act also compels the employee to take reasonable care. It may be difficult for him to appreciate what is 'reasonable'. Consider the following definition: 'The factor of reasonable care is the equation that compares the risk against the time and cost of eradicating the hazard.' This can be better understood by the following comparison.

A painter sanding down a window prior to painting is exposed to a low risk. If he uses goggles the cost is £1.20 for suitable protection and the time taken to fit the goggles two minutes. The chances of blindness are very remote and the painter rarely uses eye protection in these circumstances. Protection against this low risk costs two minutes and £1.20.

A woodwork machinist should wear goggles whilst using a circular saw. The cost of suitable goggles is £1.30 and the time to fit them two minutes. The possibility of lost eyesight is a medium risk because the machine is designed with a crown guard to prevent particles of wood from hitting the machinist. Risk does exist, however, and goggles should be worn. Protection against this medium risk costs two minutes and £1.30.

A welder in a workshop should wear, for eye protection, tinted goggles while welding. The goggles cost £1.50 and the time to fit them again two minutes. The possibility of damage to eyesight is very high because of the intense glare and light. Protection against high risk has cost two minutes and £1.50.

The factor of reasonable care is therefore expressed in terms of simple human care and consideration. In each example the time and cost are similar but the degrees of risk to sight are very different, varying from the remote to the serious. Reasonable care is the normal, sensible care taken by the average person irrespective of time or cost. In the case of welding a far greater expense is both reasonable and necessary to maintain adequate safety.

The introduction of the Health and Safety at Work etc. Act 1974 has created several areas of support for the employee as follows.

Improved information

The employer has to offer information and instruction to reduce the risk to the employee and maintain a good level of safety. This information should be available from manufacturers but must be passed on by the employer.

Improved knowledge of risk from manufacturers

Employees must be made aware, by improved literature, of risks that may exist from the design or manufacture of any goods or substance they may use. Any such article must be clearly marked with the danger defined and explained in case the employee is placed at risk.

Safety policy

Unless exempted by employing less than five persons, the employer should give the employee a personal copy of the company safety policy. This informs the employee completely of the requirements laid down by the employer to maintain adequate safety standards. Areas of responsibility are detailed so that the employee can understand clearly what is expected of him and what he can expect from the employer to his employees.

Control of the self-employed

It has always been a concern of employees that persons not under the guidance and control of the same employer, may create risk or danger at their place of work. This may not be through the direct negligence of the self-employed person but due to his lack of knowledge or understanding of what is required of him. The self-employed are controlled by the same Health and Safety at Work etc. Act 1974 and are legally and morally obliged to safeguard the health and safety of themselves, and of others affected by their activities.

Inspector's advice to employees

If, during an inspection the Inspector sees a danger to the employee, he is legally obliged to advise direct of such danger so as to eliminate any risk immediately. It has always been considered bad practice to communicate directly with an employee, so the supervisor should be told; he is then responsible for informing the offending or endangered employee himself. A diplomatic approach is required to avoid resistance or apathy from those involved. It is important for an Inspector not to overrule, criticize or create difficulties for the supervisor, although a firm guiding action is needed from the Inspector. At the completion of the inspection the Inspector usually produces a written report for the management or supervisor which should include advice previously given verbally to the employee. A copy of the report should be given to the employee or his representative.

Extension of 'accident prevention' thoughts

The original concept of 'accident prevention' is now enlarged to the idea of 'health and welfare of employees'. The term 'health' indicates consideration of activities that may not be physically hazardous but could involve environmental risks. Chemicals can be a health hazard and may cause illness or death long after the employee was exposed to the hazard. The new Act requires prevention of risk to employees of a physical or environmental nature.

In conclusion it is important to make clear what may happen to employees who fail by lack of knowledge or intention to meet their legal obligations. It is an offence by an employee to:

Fail to take care so far as reasonably practicable
Damage intentionally or use recklessly anything so that it would not maintain the requirements of the Health and Safety at Work etc. Act 1974.
Obstruct the Inspector during his execution of normal duties
Obstruct or prevent anyone from appearing before or talking to an Inspector during his investigation
Fail to observe the requirements on any Improvement Notice or Prohibition Notice issued by the Inspectorate
Give false evidence regarding practices seen or experienced
Make any false entry within a statutory register, document or notice
Pretend to be an Inspector

If an individual is proven to be guilty of any offence as an employee there can be a fine, not exceeding £1000, or imprisonment, on summary conviction.

In brief summary, the employee is required legally and morally to uphold a degree of safety that would be achieved by the normal reasonable activities of sensible work procedures.

4 People and their responsibilities

Within the construction industry there exist several kinds of employer, differing in size and structure. Building establishments fall into four different groups:

A *sole trader* runs the company business completely on his own initiative and financial support from his personal subscription. Its success is to his personal benefit and its failure is equally his own risk: he has 'unlimited liability'. All financial responsibility is borne by this individual. The sole trader is a small building businessman. This type of establishment makes up approximately 50 per cent of the total construction industry.

A *partnership* involves at least two and usually not more than twenty people who collectively finance and run the business. The support, both physical and financial, is divided between the partners. The individuals establishing such a partnership are committed to support the concern financially and physically and receive remuneration in equal ratio to their investment. Partners working on a day-to-day basis are referred to as 'working partners' whilst any financing the business, but not actually participating in the daily activities, are 'sleeping partners'.

A *private company* is owned generally by a number of known shareholders. They have the right and privilege to control the activities of the business by finance and/or participation.

A *public company* is one whose owners have elected to offer interests in the business to the public. The shareholdings are offered to the public for purchase by investors in an agreed exchange for dividends as established within the company memorandum of agreement or company constitution. In small public companies the original owner often retains a majority holding of shares; by owning more than half of the shares he keeps control and can avoid any major decisions that may be derogatory to him. The company policies are established and maintained by a board of directors on behalf of the shareholders. An annual meeting is usually held to adjust any leading personnel by voting of the shareholders.

Individual responsibilities vary according to the type of company. Although establishing lines of demarcation, one should not disregard the importance of some sections nor discredit the individuals involved in the safety of the company. Everyone has some responsibility for the standard of his own health and safety and of those affected by his/her actions. For the purpose of general safety standards companies – whether public or private – can be divided into three categories:

Large companies	–	more than £5M turnover
Medium companies	–	between £1M and £5M turnover
Small companies	–	less than £1M turnover

Safety responsibilities within a large construction organization
(more than £5M turnover)

All personnel are legally and morally responsible for the safety of all employees and must accept certain levels of responsibility. The system of control in Figure 1 (page 28) is typical.

The board of directors

The directors collectively have to establish a control upon all company personnel who enforce acceptable safety standards. An individual director should be chosen to take the responsibility at board level. This

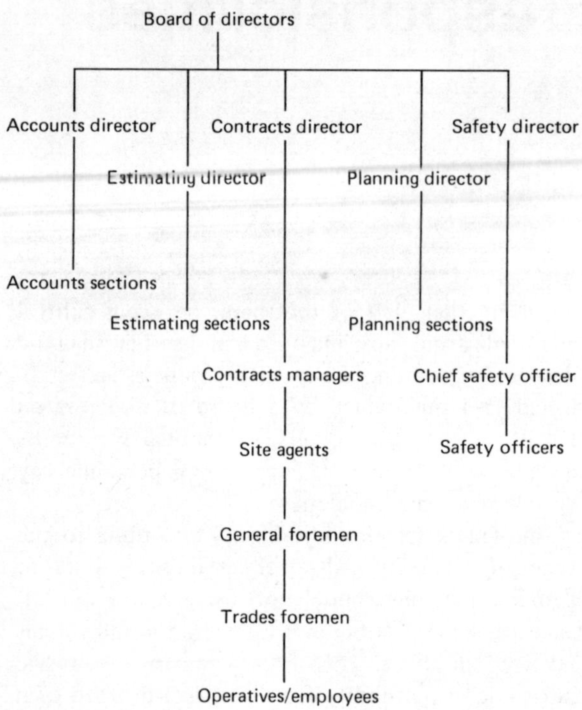

Figure 1 *Structure of a large construction organization*

director should create, maintain, and supervise the company safety. Much of this work is delegated; his ultimate duty is to report to the directors and to ensure that all management decisions are communicated. The safety director appoints a Chief Safety Officer or any other subordinate he needs to implement the company safety policy. This director also ensures that company finance is made available to the safety section to allow compliance with the best safety standards.

Chief Safety Officer

The Chief Safety Officer is appointed by, and responsible directly to, the board of directors and in particular to the director responsible for safety matters. As leader of the safety section, he is in close liaison with his subordinates, and controls all safety aspects. His principal tasks are to:

Prepare, agree with directors, and install a company safety policy

Establish a detailed safety standard, in accordance with the director's requirements, which ensures adequate protection for all employees

Maintain the correct moral and legal obligations, within the confines of both civil and criminal liabilities

Liaise with all company management sections

Provide the full range of knowledge and advice expected of a qualified Safety Officer

The Chief Safety Officer should be a member of the Institution of Industrial Safety Officers and be fully conversant with how to administer a safety policy effectively

Safety Officer

The Safety Officer, responsible to the Chief Safety Officer, is appointed by him in conjunction with the personnel department, and safety director. The terms and conditions of Safety Officer appointments are compatible with company policies. Each Safety Officer, if a team is created, has responsibilities delegated by the Chief Safety Officer. The following are responsibilities of each Safety Officer:

1 In an advisory capacity to senior management:

Recommend improvements to the existing establishment

Create and maintain an awareness of all changes in legislation

Anticipate possible hazards at new sites and recommend relevant procedures

Review the suitability, reliability and safety standard of plant for purchase or hire

Recommend the adequate provision of apparatus, protective clothing and safety equipment

Maintain a full knowledge of all legal requirements and convey details of them as necessary

Establish safe working procedures and policies

2 In a supervisory capacity to the site management team or workshop leader:

Complete site surveys regarding the company safety policy in close liaison with the site team leader or the site safety representative

Compel persons to observe all regulations concerning health, welfare, and first aid facilities

Review all periodical registers to maintain records for both company and statutory requirements

Promote and foster a clear understanding by everyone that safety is real, required and recommended

3 In a consultancy position to the company, with his higher management, his peers and subordinates:

Provide statistics and analysis of accidents, both recorded and reported, for the company and make recommendations for improvement

Establish, where possible, the cause of accident, incident or dangerous occurrence and recommend for future avoidance

Maintain a library of all past and current safety literature and circulate all information as necessary

Retain a full and accurate knowledge of all legislation

Establish and maintain schemes of safety training and advice to all levels of the company

Provide a good relationship with all professional, legal and advisory bodies and organizations related to the legal control and moral welfare of the company and its employees

Contracts Manager or Construction Manager

This person, responsible to the director in charge of contracts, is engaged by the board of directors in conjunction with the personnel officer. The Contracts Manager's task is to ensure a good flow of production and productivity upon several building sites. Each of these building contracts is organized in detail by a site team leader (site agent or general foreman). In addition to his managerial responsibilities, the Contracts Manager assumes safety responsibilities to:

Understand and implement the company safety policy and appreciate its extent of responsibility including areas of demarcation of tasks, work involvement, etc.

Be involved at the planning stage with the details of work procedures and method

Delegate a clear responsibility to all subcontractors through liaison, planning and co-ordination

Plan how best to deal with existing and future services (gas, electricity, telephone, sewage, etc.)

Decide in conjunction with service suppliers how best to provide temporary construction services

Arrange site welfare, sanitary conveniences and other statutory provisions needed on site

Organize basic fire precaution, prevention, and procedures with local fire officer and contractor's safety officer

Anticipate hazards, recommend precautions and instruct all personnel which sequence of operations to follow

Liaise with construction staff and safety officer over site procedures

Set a good example by wearing appropriate safety garments and showing an awareness of safety procedures whilst on site

Emphasize the personal advantages to be gained by good safety practices

Provide a good feedback from observations and continue to see that all statutory requirements are met on site in conjunction with the safety officer

Plant Manager

It is not uncommon for a large organization to establish a separate plant section with a Plant Manager as section leader. The co-ordination and planning of plant activities are his principal involvement within this section. However he has a job specification that varies according to each firm's structure. His safety responsibilities are to:

Ensure that all periodical tests, inspections and reports are completed

Provide a good service for each site, whereby checks upon hired plant are carried out in accordance with safety requirements. This does not involve physical maintenance because the hire company should incorporate this service into their terms of agreement. The plant officer reviews the current certificates and legal documents

See that all plant provided to site is in good condition, with all guards and other safety devices required by current legislation

Ensure, together with management, including safety officer, that all plant operators and banksmen are fully trained and competent to use their machine or apparatus

Maintain a good plant department with prompt service to site and information to management regarding defects to plant, hazards and replacements

Set up a system of maintenance to cover any deficiency in plant and to minimize the risk to operatives

Safety responsibilities within a medium-sized construction organization

(between £1M and £5M turnover)

Figure 2 provides additional information regarding the relationships of the following persons. As previously stated, all personnel are legally and morally responsible for the safety of employees. The details expressed below are typical but every company management structure is different.

The board of directors

The board is a similar collective group to that which has been discussed within a large organization, and has similar safety responsibilities. Delegation to subordinates is less as there are fewer employees. The director responsible for safety participates more in actual work activities, although his role in safety matters is similar to that in a large firm. Numerous subordinates would, however, need to be employed with responsibilities as outlined in the following sections:

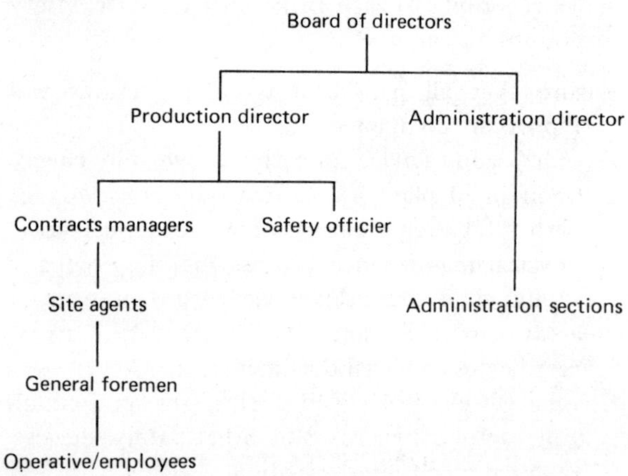

Figure 2 *Structure of a medium-sized construction organization*

The contracts manager

The contracts manager is responsible to the directors for the complete running of the sites, including personnel and procedures involved. He has several assistants but his individual responsibilities for safety needs are to:

Maintain good, safe working practices

Plan the company policy, regarding the Health and Safety at Work etc. Act, prevent injury, reduce accidents and set incentives to improve safety standards

Understand and implement in conjunction with the safety officer, current legislation and ensure the company complies with the law

Liaise closely between main contractor, subcontractor, or any other groups involved with company procedure and work

Organize a system of recording, reporting, investigating and costing any accident. Statistical analysis and control of accident trends is done with the help of the safety officer, who may be delegated to do most of this work

Encourage a rapport with all safety organizations and distribute all information about accident prevention

Arrange for acceptable expenditure on safety needs, as requested by the safety officer

Discipline, as necessary, any subordinate who has not met the company safety standards and has ignored lower level warnings

Keep all safe practices in mind while tendering, planning and building

Company safety officer

A safety officer is employed but the company, probably, does not establish a safety department. Smaller companies within this sector may find it advantageous to employ a part-time safety officer who can see to safety requirements and also another designated responsibility, maybe transport and plant or personnel. This safety officer has responsibilities similar to those of a chief safety officer but varying in size, scope and detail, and could be:

Advise how injury or hazard to personnel can be kept to a minimum

Advise on all legal changes or developments and recommend precautions
Warn about probable hazards on new contracts
Distribute literature and information regarding clothing or equipment
Advise on safety aspects of plant for purchase or hire
Inspect sites with the site team leader and produce relevant submissions
Obtain, then display the relevant and required statutory documents and see that the contents are observed
Create a good first aid procedure for each site
Promote a safety-conscious attitude throughout the company and compare his company's accident statistics with others
Record accidents, having determined the facts, so that future dangerous occurrences are kept to a minimum

Safety responsibilities within a small construction organization

(less that £1M turnover)

This group of contractors and builders are the most numerous of all. The style and details of firms' organizations vary. The reader is advised to read and understand the first section of this chapter to be familiar with the terminology of smaller company structure and organization. The principal types of small builder are the sole trader and partnership. The private company is also well represented in this group.

The following is a basis for safety structure within the company, which may be supported by external aids such as group safety practice, safety advisors or consultants. It varies according to the individual firm. The structure incorporates the safety responsibilities legally and morally imposed upon all personnel.

The manager

The manager carries direct responsibility to the company owner and acts as a leader and organizer of the firm's activities according to the owner's requirements. Among his safety responsibilities are:

Where legally required, i.e. where employing five or more persons, plan, draft, prepare and issue the safety policy and make all provisions to implement the policy
Arrange any financial allowance that may be necessary to establish acceptable safety standards
Co-ordinate with all directly employed persons and subcontractors
Make all provisions at planning, estimating or tender stages for safety
Understand and implement all current legislation
Set up a method of recording and reporting accidents, incidents or dangerous occurrences
Liaise with accident prevention organizations and distribute their literature
Organize a way of motivating the workforce towards safe practices

Safety responsibilities of personnel, regardless of company size or activities

The preceding section has dealt with the management structure, its personnel, and their safety responsibilities. The following construction employees, in any size of company, have a varying role according to individual position, status or job description.

Site agent

A site agent is generally found only in larger companies, and often is called the site team leader. He deals with day-to-day decisions in conjunction with the contracts manager. The duties of the site agent are to build correctly, economically, quickly and safely by utilizing the resources available to the best possible advantage. His responsibilities include all aspects of building work but his specific safety responsibilities are to:

Organize the site to enable work to proceed without risk to men or damage to equipment and building, including any fixed or unfixed materials
Organize the work so the site is kept tidy
Arrange the installation, maintenance and correct use of all power and services especially electrical apparatus

Provide all protective clothing and apparatus and ensure their availability and use

Co-operate with safety representatives, especially the safety officer. Particularly if there is no safety officer, close links should be made with the safety inspector, fire prevention officer and public health inspector

Check on plant, machines, equipment, powered hand-tools, hand-tools, and all other working devices with particular regard to safe practices. This requires clearly planned delegation to a subordinate

Organize and maintain a policy of correct and safe storage of materials, particularly of hazardous substances like chemicals, liquids and gases and provide adequate information to persons using any new, experimental or hazardous materials

Give adequate instructions on working procedures to all operatives, especially apprenticed and trainee workers, making sure none are exposed to risk

Keep up to date with all current legislation and statutory documents, and systematize the keeping of documents, records, reports, registers and forms that may be required by company policy or by legal obligation

Make sure, with the qualified first aid personnel, everything is available for use in emergency and that all equipment morally and legally required is at hand

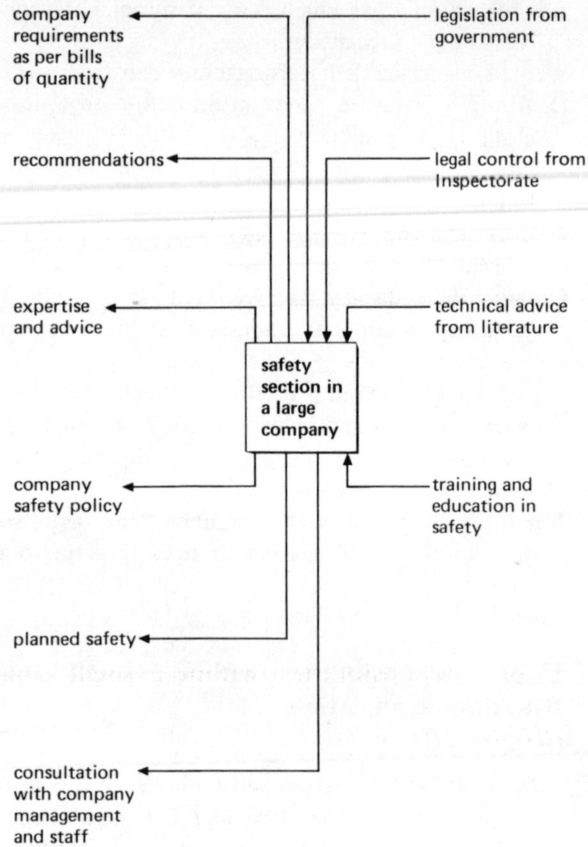

Figure 3 *The role of the safety section in a large company*

General foreman/trades foreman/ganger

These are the subordinates of the site agent and have a workload delegated by him. Each of these 'section leaders' is expected to bring management expertise to responsibilities including good building practices, procedures and economics. In addition, he is expected, in order to execute fundamental safety policies, to do the following:

Incorporate the needs of safety within all instructions about work

Discourage any bad practices, rebuke any operatives putting themselves at risk, create good safety procedures and command all proper safety activities

Ensure that all newcomers to the site, especially apprentices and trainees, take adequate precautions by providing all knowledge to them and explaining their legal and moral obligations

Restrain all operatives, both directly employed and sub-contractors' employees, from taking any risks

Understand all relevant legislation and implement it at the place of work

Report any defects in plant, machinery, equipment or apparatus to superiors (it may be quicker to to organize the repair and then report to the senior staff)

Co-operate with visiting safety advisors and share responsibility with both superiors and subordinates to create good safety practices for mutual advantage

Set a good example to others and see that all practicable precautions are adhered to

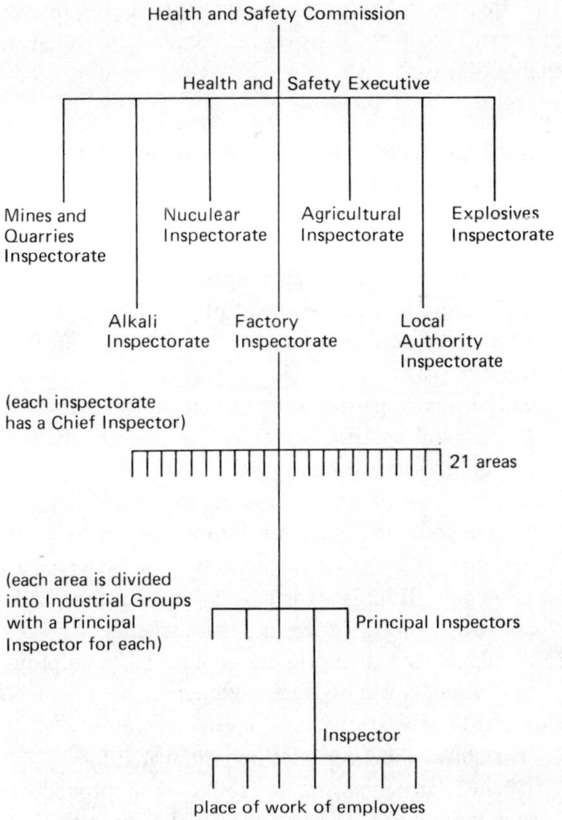

Figure 4 *Levels of control for health and safety*

Operatives

An operative's contract of work expects him to give his time, experience, expertise and knowledge to the best of his ability in return for his pay. The operative is therefore employed to do a good job, at a reasonable rate of output. His terms of safety are based upon a 'contract of employment' rather than a 'place of work'. His safety responsibilities are:

Avoid horseplay, or an action of folly or stupidity that may reduce the protection of himself or any adjacent workers

Use all suitable tools, equipment and apparatus for the job, including protective garments and equipment, maintain these in good working condition without causing any hazard, and not reduce the level of safety by improvising

Take full and proper advantage of all facilities and equipment provided for safe use

Report to superiors any defect to equipment, tools or apparatus that may be a safety risk

Guide and encourage all new employees, especially apprentices, to be safe, and make them aware of any particular hazards

Develop a personal interest in safety and offer any suggestions to reduce risks at work.

It is a requirement of the Health and Safety at Work etc. Act that *all* employees must by statutory law:

Co-operate as far as possible with the employer to achieve the requirements of the legislation

Never intentionally damage or use recklessly anything provided for safety

Take care, so far as reasonably practicable, to maintain the employee's own safety and that of anyone else who may be at risk because of his work.

Safety responsibilities of people outside the construction organization

Safety groups

A safety group is a team made up from company members, consultants or individuals who work together for good safety. Areas organize a function group of interested people who promote safety by lectures, seminars and conferences. Finance is usually secured by subscription. A good local safety group might be expected to.

Advise and promote safety in local industry by way of meetings and functions

Establish an awareness that safety needs to be successful

Circulate all knowledge of possible safety procedures

Communicate between the Health and Safety Executive and local industry to reduce the gap between top authority and the workforce

Liaise throughout local industry with personnel to mutual advantage

Safety consultant/advisor

Where several small firms find it uneconomical and impracticable to employ a full-time safety officer they club together to employ a group safety consultant. Generally any firm with fewer than one hundred operatives finds this scheme useful and economical. The consultant acts as a safety officer to each firm and, on request, advises on current site activities. There is of course no limit to the service offered nor any demarcation of size, type or scope of firm that can employ a safety consultant. The responsibilities may vary from firm to firm. The following functions would be expected from a good safety consultancy group either collectively or from each of its individuals:

Give advice on all safety matters both at site and management level

Promote safety at site level by good periodic checks on plant, equipment, and procedures and organizing records of checks and tests

Promote safety at management level by good reporting, and giving advice on matters such as safety policy, training and fault diagnosis

Regularly inspect safety standards, admonishing bad safety trends and commending good practices

Investigate accidents and minor incidents and reporting accordingly

Co-operate with authorities and controlling powers

Maintain a good knowledge and understanding of current safety procedures

The Safety Inspector

After the Robens Report of 1972, which concluded a deep-searching review of all industries, vast changes were developed in the statutory controls of working activities. By the Health and Safety at Work etc. Act the government established a new organization with enough financial backing, legal power and absolute determination to reduce the carnage within industry, particularly construction.

A system of control was established (Figure 4, page 33) which aimed at reducing death and serious injury both by physical means and health hazard controls. The Health and Safety Commission controls the work of the Health and Safety Executive. By simple line control this executive has established twenty-one inspectorate areas across the country. Within these areas the Safety Inspectors operate, maintaining acceptable levels of safety at work. The Inspector has new and stronger legal powers. The powers of the Inspector, detailed in the Act, enable him to:

Enter any premises, at any reasonable time, to fulfil the needs of his duty. Where, in his opinion, a situation may be dangerous an entry can be made at *any* time. If obstruction is anticipated he may be accompanied by a constable

May bring with him any apparatus, equipment or materials needed to fulfil his inspection. This may require the help of a qualified professional assistant

Make any investigation or examination as necessary. This includes noise meter reading or checks for air contamination or any pollution

Take photographs and measurements that may be needed to record fully the findings of his visit

Take samples of substances: material, chemical or gas

Remove any findings from the investigation, and take possession of them as legal evidence

Take factual and accurate statements from employees, which could be legal evidence

Check that any statutory documents comply with laws about their completion and availability

Carry out any reasonable activity or procedure that would be of assistance to the investigation

Having carried out a thorough review of the workplace, the Inspector can assess the situation and take one of the following actions:

He can leave the place of work content and satisfied that all is well, within the legal and moral requirements of all parties.

If the Inspector forms the opinion that a hazard exists that does not cause imminent or immediate danger, he issues an Improvement Notice (page 48). Improvement Notices are issued when defects exist but do not cause serious danger, for example failure to display statutory notices, or providing a cold water supply but no hot water. The Improvement Notice gives the employer a specific date by which to correct the error. The offending employer can appeal against this action; in this case, the Notice is suspended until an industrial tribunal assesses the situation.

If the Inspector forms the opinion that an

Improvement Notice has not been complied with, or that a hazard exists of imminent danger then a Prohibition Notice is enforced. Typical examples might be failure to have secure and adequate scaffolding, or failure to have properly guarded machines. The Prohibition Notice has the direct effect of stopping work until the hazard is eradicated.

It is a legal offence (section 33 of the Act) to imitate or pretend to be an Inspector, with a heavy fine on conviction.

The Inspector's lot is not a happy one and is aggravated by numerous ways of boycotting his activities. But he is still a human and needs to be respected for his knowledge, academic and practical skill. He has heavy moral and legal responsibilities to carry out without prejudice. The Inspector's responsibilities are generally those listed below, although they vary and may be extended according to the industry or style of the Inspector:

Be fully conversant with current legislation

Be consistent with the interpretation and implementation of the legal requirements, and make regular inspections and visits

Be firm and fair to the people who are subject to his authority

Avoid bias or favour and keep a simple but direct approach that can be understood and accepted by those being inspected

Enforce the law, and avoid any varied interpretation

Offer advice on, and explain the meaning of, legal documents to those who may be unsure of what is required of them

Liaise with the representatives of companies, especially safety officers or site team leaders, and be prepared to advise on safety activities. Commend all good procedures and condemn all hazardous working, whilst avoiding a too-close relationship that may undermine his authority or make industry complacent about the Inspectorate

Confirm all actions by report, certificate, instruction or letter. This eliminates any lack of understanding caused by poor communication

Avoid an insincere or overpowering manner with those beneath his power

It is difficult to list all the responsibilities of a perfect Inspector, but he is always especially concerned to eliminate unsafe practices and the keeping of false records.

Section 39 of the Act permits the Inspector to represent the Executive at a magistrates' court. Except in Scotland, the Inspector assumes the position of solicitor or counsel in magistrates' court.

The competent person

Several references are made to the 'competent person', who is given responsibility for checking and reporting the stability and safety of certain parts of the workplace. There does not appear to be any legal definition of a competent person, or of the individual standards required from the person doing this work. The checking and approving of scaffolds, excavations and similar hazardous places of work, is delegated to a competent person. He needs to be of sound character, conscientious, accurate and unbiassed. A competent person needs to have practical knowledge of the workplace and should, ideally, have worked with the machine, apparatus or plant. In addition, he needs a sound theoretical understanding of them and fully to appreciate the implications of his inspection. A further requirement of the competent person is that he must be able to detect a weakness and relate its importance to the function, stability and safety of the equipment machine.

A simple checklist of responsibilities of the competent person includes the following:

Know the workplace or equipment to be inspected, practically and theoretically

Appreciate the level of hazard and decide on improvements that are legally acceptable, but also maintain a good level of work for the company

Avoid being unfairly strongheaded at an inspection

Take a logical approach to the jobs required of him

Keep accurate records and complete registers of all inspections

Never avoid, nor miss any statutory inspection, nor reduce the standard of inspection

First aider

As a statutory requirement of the Construction (Health and Welfare) Regulations 1966 as amended in 1974, there must be a good standard of first aid available on all sites where any contractor employs more than five persons. (For a more detailed appraisal see page 175.) A qualified first aid person must be available to deal with all needs and requirements.

The definition of first aid is: assistance given to an injured person before professional aid arrives. It must be emphasized that the initial aid given may well be crucial to the future well-being of the injured person. Where no further assistance is needed the first aider provides the only necessary treatment. Regardless of the level or intensity of first aid required, the person doing such a job needs to:

Be fully conversant with first aid procedures dealing with construction injuries

Understand the requirements of first aid boxes as defined within the Construction (Health and Welfare) Regulations 1966 and maintain all stocks as applicable at the workplace

As a minimum requirement, hold a certificate, not older than three years, of either the St John Ambulance Association or the British Red Cross or any similar society (no one under the age of fifteen years can be an approved holder for construction purposes)

Be available at all the times required by legislation, according to the workplace

Offer advice to all other employees regarding first aid

Designers, manufacturers and suppliers

The Act requires a greater regard for safety in all areas of work, both with regard to completed construction and to equipment and substances used. Within the construction industry, this gives a responsibility upon all design personnel, architectural or structural designers and planners, to consider their moral obligations:

Consider if the building processes needed to complete the proposed work provide acceptable working conditions

Make allowances for permanent repairs and maintenance by incorporating anchorage for scaffolds, cradles etc.

Plan for adequate structural stability by calculated loading, bracing and supporting as necessary

Allow for human limitations and design with the particular hazards of infants, the elderly and the disabled in mind

Anticipate and reduce pollution where sewage, waste disposal or products of combustion may cause safety hazards

When designing a machine or functional piece of apparatus, consider its safety whilst retaining its practical use

Complete all necessary research to eliminate risks to the person using machinery, apparatus or the completed construction

As a supplier, issue literature and clear information so the receiver is fully aware of any hazard

The self-employed

A self-employed person may be:

Professional, e.g. construction surveyor, accountant

Businessman, e.g. director, trader

Private home worker, e.g. representative, salesman

Subcontractor, e.g. specialist designer, specialist installer or fitter

The last of these is of most concern to the construction industry. The subcontractor is employed under a contract, without this specifically being a full contract of employment. The subcontractor, whether providing materials and labour or labour only, is considered to be self-employed.

The legal and moral requirements expected from the self-employed within construction work are:

Avoid any risk to himself or anyone else during work activities

Take all precautions so far as reasonably practicable

Give full information to those who may be affected by work activities to minimize their exposure to risk

Advise and warn any public persons (third party) who may be exposed to a dangerous situation resulting from self-employed working

As well as this list of specific subcontractor responsibilities, there are a great many others which may be borne by the main contractor or by the subcontractor:

Check that a clear demarcation of responsibilities exists with the main contractor
Agree with the main contractor that facilities are available for shared use, e.g. scaffold, welfare, storage, fire precautions
Ensure that all statutory notices are provided either by the main contractor or by the subcontractor
Establish a relationship with the safety officer for the main contractor and ensure their policy of safety is understood and implemented. Where the subcontractor employs more than five operatives he will need his own safety policy (see page 22)
Accept responsibility for his own employees and instruct, control, organize and employ them in the same way as an employer (as defined on pages 19-21)
Where no main contractor is in a responsible position, share the overall legal responsibilities
Provide adequate first aid and welfare facilities (there may be exemptions where agreement has been made (Form 2202) between all parties involved)
Where working with plant as its installer, user or hirer to ensure that all relevant laws are obeyed. Certain hire situations need a shared responsibility with the hire company according to the conditions of hire

Colleges and training establishments

'Learning continues for the duration of a lifetime.' The length of a lifetime may depend on the individual care taken by employers, superiors, peers or subordinates. Colleges (establishments of further education) or training centres have an obligation to the students. Under Section 2(C) of the Act the employer provides training for his employees. Training for craft and management has been successfully available for years but the new legal requirements ensure that safety is an integral part of the studies. The responsibility for producing a good workforce lies with both administration authorities and the lecturer and they should:

Give students a complete and thorough knowledge
Provide typical work situations and incorporate full safety requirements
Set a good example to all who are influenced by activities at college
Maintain good working practices by fulfilling statutory obligations
Apply a safety code of practice that simulates typical work requirements
Create student participation to improve confidence and motivate enthusiasm
Compliment good practices and condemn malpractices, showing methods of improvement
Give time, patience, guidance and knowledge to all students

Students

Students who attend colleges or training establishments to further their knowledge of construction may be on day release, block release or full-time study. All work must be done in accordance with the safety policy of the college and must at all times comply with the minimum legal safety needs. The responsibilities of students are generally (these may vary in different trades and colleges):

Work in a safe manner, in accordance with the instructions given
Work in the same way as a normal employee, taking the legal and moral responsibilities of reasonable care (see also page 24)
Obey the instructions given by the class lecturer and respect his position
Use only equipment and machines within the capabilities of one's experience and ability. Special provision is made for certain machines to be used by those under the age of eighteen
Provide the correct clothing and protective equipment for safe working
Observe all safety procedures that are established within the college
Act and work well and make every use of the facilities available
Take all reasonable care

5 Documents

The documents referred to in this chapter are complementary to the legislation and statutory instructions discussed in chapter 20 (pages 166-76).

A procedure of forms, certificates and notices to confirm or check actions is a necessary back-up to the legal system. The following booklets and forms are reviewed in conjunction with the legal requirements attached to various work situations. They are not all necessary on all sites, but company safety representatives should understand the form of each document.

Accident book (Form BI 510) (Figure 5)

This booklet should be available at any workplace employing ten or more persons at the same time. It is a supporting document of the Social Security Act 1975 and is designed, prepared and used with the approval of the Secretary of State for Social Services. The most important features and implications of this booklet are:

An accident should be recorded as soon as practicable after it occurs

Details should be completed by the injured person or his/her *bona fide* representative

The accident book should be retained for at least three years after the last entry

Various advisory notes are included for claims under the National Insurance Benefit Scheme

Record of inspections of scaffolds (Form 91 Part 1 Section A) (Figure 6)

The booklet supports the legislation of the Factories Act 1961 and the Construction (Working Places) Regulations 1966. Its function is to maintain a record of inspections to scaffolds (see page 40 for details). The document advises that:

The contractor's name and address must be clearly shown on the report

All reports in the booklet should be retained on site, or after completion, for a period of two years at the employer's office, and be available for inspection by HM Inspector of Factories

The inspections must be carried out every seven days, and after inclement weather, or after any structural change

If work lasts no longer than six weeks the employer is exempted from this report procedure. However in such cases the site leader confirms his inspection to the employer in writing

Record of weekly thorough examination of excavations (Form 91 Part 1 Section B) (Figure 7)

A booklet of forms prescribed under the legislation (Regulation 9) of the Construction (General Provisions) Regulations 1961. The reports are to confirm the completion of inspections, and notes for guidance include:

The results of examinations are to be retained on the relevant site or, after completion of work, at the office of the employer, for at least two years. During this time the register of forms needs to be available for inspection by HM Inspector of Factories

The requirement of the examination is that any part of excavations, earthwork, etc., shall be inspected at least once every day of work. The last two metres of a working trench should be inspected before each working shift

See Instructions on pages 1 and 2

ACCIDENT BOOK, as approved by the Secretary of State for Social Services for the purposes of the SOCIAL SECURITY ACT, 1975

Full Name, Address and Occupation of Injured Person (1)	Signature of injured person or other person making this Entry (If the entry is made by some person acting on behalf of the employee, the address and occupation of such person must also be given) (2)	Date when Entry made (3)	Date and time of Accident (4)	Room or Place in which Accident happened (5)	Cause and Nature of Injury (State clearly the work or process being performed at the time of the accident) (6)

Figure 5 *Accident book*

Name or title of employer or contractor ______________________

Address of site ______________________

Work commenced - Date ______________________

Factories Act 1961

Construction (Working Places) Regulations 1966

SECTION A

SCAFFOLD INSPECTIONS

Reports of results of inspections under Regulation 22 of scaffolds, including boatswain's chairs, cages, skips and similar plant or equipment (and plant or equipment used for the purposes thereof)

Location and description of scaffold, etc. and other plant or equipment inspected (1)	Date of inspection (2)	Result of inspection. State whether in good order (3)	Signature (or, in case where signature is not legally required, name) of person who made the inspection (4)

SPECIMEN

SHORT CHECK LIST*—AT EACH INSPECTION CHECK THAT YOUR SCAFFOLDING DOES NOT HAVE THESE FAULTS

		Week 1	2	3	4
FOOTINGS	Soft and uneven No base plates No sole boards Undermined				
STANDARDS	Not plumb Jointed at same height Wrong spacing Damaged				
LEDGERS	Not level Joint in same bays Loose Damaged				

		Week 1	2	3	4
BRACING 'Facade and ledger'	Some missing Loose Wrong fittings				
PUTLOGS and TRANSOMS	Wrongly spaced Loose Wrongly supported				
COUPLINGS	Wrong fitting Loose Damaged No check couplers				
BRIDLES	Wrong spacing Wrong couplings Weak support				

		Week 1	2	3	4
TIES	Some missing Loose				
BOARDING	Bad boards Trap boards Incomplete Insufficient supports				
GUARD RAILS & TOE BOARDS	Wrong height Loose Some missing				
LADDERS	Damaged Insufficient length Not tied				

**This check list is not part of the report required by Regulation 22*

Figure 6 *Form for inspection of scaffolds*

9

Name or title of Employer or Contractor ....

Address of Site ....

Work commenced Date ....

Excavations, Shafts, Earthworks, Tunnels, Cofferdams and Caissons

Factories Act 1961 — Section B

Reports of results of weekly thorough examinations
in pursuance of regulation 9 (*Excavations, etc.*)
and regulation 18 (*Cofferdams, etc.*) *of*
the Construction (*General Provisions*)
Regulations 1961

Description or location 1	Date of examination 2	Result of thorough examination State whether in good order 3	Signature (or, in case where signature is not legally required, name) of person who made the inspection 4

SPECIMEN

Figure 7 *Form for weekly examinations of excavations, etc.*

No person shall be employed within a trench or excavation that has not been inspected

The employer is exempt if he reasonably expects the work to last less than six weeks. In that case the site leader's written confirmation of inspection to his employer is satisfactory for the legislation requirements

Records of weekly inspections, examinations and special tests of lifting appliances (Form 91 Part 1 Sections C-F)

This compilation of forms applies to several situations, all required by the Construction (Lifting Operations) Regulations 1961. The book is divided into sections covering lifting appliances (cranes, hoist, gin wheels, etc.), cranes (anchorage and ballasting tests), automatic safe load indicators, hoists used for carrying passengers. The following summarizes the notes for guidance included within the register.

Section C – lifting appliances (Figure 8)

To comply, an inspection should be made once a week, either by a competent person or driver.

The safe loading indicator must be inspected at least weekly when the appliance is in use.

Exemption is permitted where the work is not expected to last more than six weeks, in which case a written confirmation to the employer is sufficient.

Section D – anchorage and stability (Figure 9)

Inspection must be made after each erection and after substantial adjustment or movement.

Tests must be completed before use.

Section E – safe load indicators

Mobile jib cranes should be tested every time they are dismantled in part, or in whole, or after any adjustment which may affect the indicator's operations. (A 'mobile crane' is a self-propelled and not a rail-mounted type.)

Tests to other jib cranes shall be made every time the crane is erected.

Tests to be completed before use.

Only in the case of mobile cranes is testing by the driver acceptable.

Section F – passenger hoists

A hoist must not be used unless tested after erection or following any height adjustment.

Testing must be carried out by a competent person.

Record of reports on examinations of lifting appliances, hoists, chains, rope and lifting gear (Form 91, Part 2 Sections G-K)

This book of registers is a compendium of reports required by the controls of the Factories Act 1961 and the Construction (Lifting Operations) Regulations 1961. It is a continuation of Form 91 Part 1 and consists of several similar requirements of completion, retention and availability to HM Inspectors. The requirements within the register can be summarized as follows:

Section G – lifting appliances (except hoists)

A thorough examination is required every fourteen months or after substantial repair/alteration (Regulation 28).

A signed report must be available and completed within twenty-eight days of inspection.

If the report states that the lifting appliance cannot be used safely without repair or alteration, a copy must be forwarded to HM Inspector of Factories.

Certificates of test and examination must be in the prescribed Form 96 for cranes and Form 80 for other lifting appliances.

Section H – hoists

Hoists must not be used unless thoroughly examined, at least once within the previous six months.

A report of the inspection must be made within twenty-eight days.

If the report shows that the lifting appliance cannot be used safely without repair or alteration a copy should be forwarded to HM Inspector of Factories.

2?

Name or title of employer or contractor ______________________

Address of site ______________________

Work commenced. Date ______________________

FACTORIES ACT 1961

LIFTING APPLIANCES

SECTION C

Aerial cableways, aerial ropeways, crabs, cranes, draglines, excavators, gin wheels, hoists, overhead runways, piling frames, pulley blocks, sheer legs, winches

REPORTS OF RESULTS OF WEEKLY INSPECTIONS

Form prescribed by the Secretary of State in pursuance of regulations 10 and 30 of the Construction (Lifting Operations) Regulations 1961

Description of lifting appliance and means of identification (1)	Date of inspection (2)	Result of inspection (including all working gear and anchoring or fixing plant or gear, and where required the automatic safe load indicator and the derricking interlock) State whether in good order (3)	Signature (or, in case where signature is not legally required, name) of person who made the inspection (4)

SPECIMEN

Figure 8 *Form for weekly inspections of lifting gear*

28

Name or title of employer or contractor ....

FACTORIES ACT 1961

CRANES

SECTION D

Address of site ....

REPORTS AND RESULTS OF ANCHORING OR BALLASTING TESTS

Form prescribed by the Secretary of State in pursuance of regulation 19 of the Construction (Lifting Operations) Regulations 1961

Work commenced. Date ....

Description of crane and means of identification	Date of test	Tests applied (3)			Safe working loads as ballasted (4)	Signature of person who made the test
(1)	(2)	Load imposed (tons)	Radius of jib (feet)	Anchorage tested	Load (tons)	(5)

Figure 9 *Form for reporting tests on the anchorage of cranes*

Certificates of test for new hoists or those which have been altered should be produced in Form F75 for hoists and Form F91 for passenger hoists.

Section J – chains, ropes and lifting gear

All chains, ropes or similar items for lifting gear must have been inspected at least once within the previous six months (Regulation 40).

Apparatus not in regular use can be inspected as is necessary.

Similar inspection is required of any apparatus which has been substantially altered, repaired or lengthened.

Certificates of test are to be supplied on Form F87 for wire ropes and Form F97 for chains and lifting gear.

Section K – reports of annealing or approved heat treatment

This incorporates a detailed analysis of those instances in which chains, etc., affected by heat, cannot be used. Regulation 41 details the exact requirements and limitations for practical work of such affected equipment.

Use of lead paint in connection with buildings (Form 92)

Required by the Factories Act 1961, this register needs to be completed with a list of persons employed, and of painting contracts in progress. The purpose of this document is to keep a check on persons who run a risk of lead poisoning. The instructions within the register are briefly as follows:

It is a requirement of the Factories Act 1961 (section 129) that a register be kept containing details of individuals and the nature of work undertaken

The list of contracts and work involved must be maintained unless the HM Inspector of Factories can get this information through other commercial sources

Every case of lead poisoning must be reported to HM Inspector on Form F41.

General register (building operations etc.) Form 36

This register is a complete record of young persons, accidents and dangerous occurrences, cases of poisoning or disease. (Duplicate forms must be kept for young persons and accidents.) It is a legally binding document under the requirements of the Factories Act 1961. Below is a summary of the essential requirements:

Any fatal accident, or one in which an employee is disabled or absent from work for more than three days, shall be recorded in Part 3 of the register. A copy of the report (Form 43) shall be sent to HM Inspector of Factories (Figures 14 and 15, pages 52-3)

All cases of poisoning and disease must be recorded in Part 4 and a copy sent to HM Inspector, regardless of whether the employee is absent for three days

Employment of young persons requires that (*a*) notification to the careers office of the Youth Employment Centre is sent within seven days of commencement; and (*b*) the young person is fit and without any defect which may affect his/her stability of safety at work

Accidents and dangerous occurrences require that (*a*) notice of death or disablement be given (as previously described); (*b*) in accidents leading to disablement after which the victim dies, a written notice must follow as soon as possible after death; and (*c*) certain dangerous occurrences must be reported regardless of disablement

Cases of poisoning or disease must be recorded on Form 41 regardless of disablement, or absence from work for more than three days

Register for purposes of Abrasive Wheels Regulations 1970 (Form F 2346)

A requirement of the Factories Act 1961 and the Abrasive Wheels Regulations 1970, this register should be an accurate record of all personnel who are suitably qualified and appointed to mount abrasive wheels. The advisory notes shown in the booklet prescribe:

Department of Employment F 2404

FACTORIES ACT 1961, EMPLOYMENT MEDICAL ADVISORY SERVICE ACT 1972

NOTICE OF TAKING INTO EMPLOYMENT OR TRANSFERENCE OF A YOUNG PERSON

Section 119A of the Factories Act 1961 requires an employer not later than seven days after taking a young person under the age of 18 into employment to work in premises or on a process or operation subject to the Factories Act 1961 or transferring a young person to such work from work not subject to that Act, to send a written notice to the local Careers Office.

NAME OF OCCUPIER ....

ADDRESS OF FACTORY OR PLACE OF WORK (*If construction industry and the young person has been taken into employment, or transferred, to work on a particular site, the address of SITE should be given*).

....

....

....

DATE OF TAKING INTO EMPLOYMENT/TRANSFERENCE ....
(*delete inappropriate item*)

NATURE OF WORK TO BE DONE BY YOUNG PERSON ....

....

....

Please give the following information so far as it is known:—

SURNAME OF YOUNG PERSON (*capitals*) ....

CHRISTIAN NAME (*OR FORENAME*) ....

ADDRESS ....

....

....

DATE OF BIRTH ....

NAME AND ADDRESS OF LAST SCHOOL ATTENDED ....

....

....

Signature .... Date ....

Position in firm ....

SPECIMEN

Department of Employment F 2404

FACTORIES ACT 1961, EMPLOYMENT MEDICAL ADVISORY SERVICE ACT 1972

NOTICE OF TAKING INTO EMPLOYMENT OR TRANSFERENCE OF A YOUNG PERSON

Section 119A of the Factories Act 1961 requires an employer not later than seven days after taking a young person under the age of 18 into employment to work in premises or on a process or operation subject to the Factories Act 1961 or transferring a young person to such work from work not subject to that Act, to send a written notice to the local Careers Office.

NAME OF OCCUPIER ....

ADDRESS OF FACTORY OR PLACE OF WORK (*If construction industry and the young person has been taken into employment, or transferred, to work on a particular site, the address of SITE should be given*).

....

....

....

DATE OF TAKING INTO EMPLOYMENT/TRANSFERENCE ....
(*delete inappropriate item*)

NATURE OF WORK TO BE DONE BY YOUNG PERSON ....

....

....

Please give the following information so far as it is known:—

SURNAME OF YOUNG PERSON (*capitals*) ....

CHRISTIAN NAME (*OR FORENAME*) ....

ADDRESS ....

....

....

DATE OF BIRTH ....

NAME AND ADDRESS OF LAST SCHOOL ATTENDED ....

....

....

Signature .... Date ....

Position in firm ....

SPECIMEN

Figure 10 *Notice of employment or transference of a young person*

Only appointed persons suitably instructed can mount the prescribed abrasive wheels

The appointed person must show a valid certificate of competence, which details the classification of wheels involved

Safety policy

In companies employing five or more persons there is a legal obligation to prepare and issue a safety policy. This has been dealt with in detail on page 22 and an example is shown in Figure 11 below. The policy must give a clear indication of the company's intentions regarding the safety of all operatives. It should be presented in a simple and precise manner, avoiding ambiguity.

Improvement Notices (Figure 12)

Reference has been made to this document (page 34) as a responsibility of the Safety Inspector. It clearly shows that the powers of the Inspectors

R & D Building Company

Pararad Road, Roselip

SAFETY POLICY

It is Company policy to afford every employee a safe and healthy place in which to work. To this end very reasonable effort will be made in the interests of accident prevention, fire prevention and health preservation by:-

1. The employment of a qualified Safety Officer directly responsible to the Contracts Director.
2. Regular visits by a Safety Consulting Officer to all places of work to make sure that the safest practicable working conditions are maintained for all employees at all times.
3. Charging line Management with responsibility for the safe operation of Plant, Site, Workshops and Offices.
4. Take all practical steps to guard any Plant or Machinery that might cause personal injuries.
5. Ensuring that accidents, fire and disaster arrangements are made and that everyone understands them.
6. Insisting on the use of protective clothing and equipment required by statutory law or wherever the Company deems it necessary.
7. Training of employees to work safely, to understand that this is to his advantage, that he must co-operate in working safely and that discipline is sometimes necessary.
8. It is the responsibility of line Management and operatives at all levels throughout the Company to implement this policy and to strive at all times to achieve the Company's safety objectives.

Contracts Director

I confirm receipt of the Company Safety Policy

Signed Date

To be returned within 7 days

Figure 11 *Company safety policy*

HEALTH AND SAFETY EXECUTIVE
Health and Safety at Work etc. Act 1974, Sections 21, 23, and 24

IMPROVEMENT NOTICE

Serial No. I

Name and address (See Section 46)	To ....
(a) Delete as necessary	*(a)* Trading as ....
(b) Inspector's full name	**I** *(b)* .... one of *(c)* ....
(c) Inspector's official designation	of *(d)* .... Tel no. ....
(d) Official address	hereby give you notice that I am of the opinion that at *(e)* ....
(e) Location of premises or place and activity	you, as *(a)* an employer/a self employed person/a person wholly or partly in control of the premises. *(f)* ....
(f) Other specified capacity	*(a)* are contravening/have contravened in circumstances that make it likely that the contravention will continue or be repeated ....
(g) Provisions contravened	*(g)* .... The reasons for my said opinion are:– .... and I hereby require you to remedy the said contraventions or, as the case may be, the matters occasioning them by
(h) Date	*(h)* .... *(a)* in the manner stated in the attached schedule which forms part of the notice. Signature .... Date .... Being an inspector appointed by an instrument in writing made pursuant to Section 19 of the said Act and entitled to issue this notice. *(a)* An improvement notice is also being served on .... of .... related to the matters contained in this notice.

SPECIMEN

LP 1
Dd 347139 5000 Pads 2/75 COH

NOTES

1 Failure to comply with an Improvement Notice is an offence as provided by Section 33 of this Act and renders the offender liable to a fine not exceeding £400 on summary conviction or to an unlimited fine on conviction on indictment and a further fine of not exceeding £50 per day if the offence is continued.

2 An Inspector has power to withdraw a notice or to extend the period specified in the notice, before the end of the period specified in it. You should apply to the Inspector who has issued the notice if you wish him to consider this, but you must do so before the end of the period given in it. *(Such an application is not an appeal against this notice.)*

3 The issue of this notice does not relieve you of any legal liability resting upon you for failure to comply with any provision of this or any other enactment, before or after the issue of this notice.

4 **Your attention is drawn to the provision for appeal against this notice to an Industrial Tribunal.** Details of the method of making an appeal are given below *(see also Section 24 of the Health and Safety at Work etc. Act 1974).*

(a) Appeal can be entered against this notice to an Industrial Tribunal. The appeal should be sent to:–

(for England and Wales) The Secretary of the Tribunals
Central Office of the Industrial Tribunals
93 Ebury Bridge Road LONDON SW1W 8RE

(for Scotland) The Secretary of the Tribunals
Central Office of the Industrial Tribunals
Saint Andrew House
141 West Nile Street GLASGOW G1 2RU

(b) The appeal must be commenced by sending in writing to the Secretary of the Tribunals a notice containing the following particulars:–

(1) The name of the appellant and his address for the service of documents;

(2) The date of the notice or notices appealed against and the address of the premises or place concerned;

(3) The name and address *(as shown on the notice)* of the respondent;

(4) Particulars of the requirements or directions appealed against;

and (5) The grounds of the appeal.

A form which may be used for appeal is attached.

(c) **Time limit for appeal**

A notice of appeal must be sent to the Secretary of the Tribunals within 21 days from the date of service on the appellant of the notice or notices appealed against, or within such further period as the tribunal considers reasonable in a case where it is satisfied that it was not reasonably practicable for the notice of appeal to be presented within the period of 21 days. If posted, the appeal should be sent by recorded delivery.

(d) The entering of an appeal suspends the Improvement Notice until the appeal has been determined, but does not automatically alter the date given in this notice by which the matters contained in it must be remedied.

(e) The rules for the hearing of an appeal are given in:

The Industrial Tribunals (Improvement and Prohibition Notices Appeals) (S1 1974 No. 1925) for England and Wales.

and The Industrial Tribunals (Improvement and Prohibition Notices Appeals) (S1 1974 No. 1926) for Scotland.

Figure 12 *Health and Safety Executive Improvement Notice*

HEALTH AND SAFETY EXECUTIVE
Health and Safety at Work etc. Act 1974, Sections 22—24

PROHIBITION NOTICE

Serial No. **P**

Name and address (See Section 46)	To ....
	....
(a) Delete as necessary	*(a)* Trading as ....
(b) Inspector's full name	**I** *(b)* ....
(c) Inspector's official designation	one of *(c)* ....
	of *(d)*..
(d) Official address	.... tel no. ....
	hereby give you notice that I am of the opinion that the following activities,
	namely:— ....
	....
	....
	which are *(a)* being carried on by you/about to be carried on by you/under your control
(e) Location of activity	at *(e)*..
	involve, or will involve *(a)* a risk/an imminent risk, of serious personal injury. I am further of the opinion that the said matters involve contraventions of the following statutory provisions:—
	....
	....
	....
	because ....
	....
	....
	and I hereby direct that the said activities shall not be carried on by you or under your control *(a)* immediately/after
(f) Date	*(f)* ....
	unless the said contraventions and matters included in the schedule, which forms part of this notice, have been remedied.
	Signature .... Date ....
LP 2	being an inspector appointed by an instrument in writing made pursuant to Section 19 of the said Act and entitled to issue this notice.

Dd 347139 5000 Pads 2/75 COH

SPECIMEN

NOTES

1 Failure to comply with a Prohibition Notice is an offence as provided by Section 33 of this Act and renders the offender liable to a fine not exceeding £400 on summary conviction or to an unlimited fine or to imprisonment for a term not exceeding two years or both on conviction on indictment and a further fine of not exceeding £50 per day if the offence is continued.

2 An Inspector has power to withdraw a notice or to extend the period specified in the notice, before the end of the period specified in it. You should apply to the Inspector who has issued the notice if you wish him to consider this, but you must do so before the end of the period given in it. *(Such an application is not an appeal against this notice.)*

3 The issue of this Notice does not relieve you of any legal liability resting upon you for failure to comply with any provision of this or any other enactment, before or after the issue of this notice.

4 **Your attention is drawn to the provision for appeal against the notice to an Industrial Tribunal.** Details of the method of making an appeal are given below *(see also Section 24 of the Health and Safety at Work etc. Act 1974).*

(a) Appeal can be entered against this notice to an Industrial Tribunal. The appeal should be sent to:—

(for England and Wales) The Secretary of the Tribunals
Central Office of the Industrial Tribunals
93 Ebury Bridge Road LONDON SW1W 8RE

(for Scotland) The Secretary of the Tribunals
Central Office of the Industrial Tribunals
Saint Andrew House,
141 West Nile Street GLASGOW G1 2RU

(b) The appeal must be commenced by sending in writing to the Secretary of the Tribunals a notice containing the following particulars:—
(1) The name of the appellant and his address for the service of documents;
(2) The date of the notice or notices appealed against and the address of the premises or place concerned;
(3) The name and address *(as shown on the notice)* of the respondent;
(4) Particulars of the requirements or directions appealed against;
and (5) The grounds of the appeal.
A form which may be used for appeal is attached.

(c) **Time limit for appeal**

A notice of appeal must be sent to the Secretary of the Tribunals within 21 days from the date of service on the appellant of the notice or notices appealed against, or within such further period as the tribunal considers reasonable in a case where it is satisfied that it was not reasonably practicable for the notice of appeal to be presented within the period of 21 days. If posted the appeal should be sent by recorded delivery.

(d) The entering of an appeal does not have the effect of suspending this notice. Application can be made for the suspension of the notice to the Secretary of the Tribunals, but the notice continues in force until a Tribunal otherwise directs. An application for suspension of the notice must be in writing and must set out:—

(a) The case number of the appeal, if known, or particulars sufficient to identify it and

(b) The grounds on which the application is made. It may accompany the appeal.

(e) The rules for the hearing of an appeal are given in:—

The Industrial Tribunals (Improvement and Prohibition Notices Appeals) Regulations 1974 (S1 1974 No. 1925) for England and Wales.

and

The Industrial Tribunals (Improvement and Prohibition Notices Appeals) (Scotland) Regulations 1974 (S1 1974 No. 1926) for Scotland.

Figure 13 *Health and Safety Executive Prohibition Notice*

are legally binding. The supplementary notes included in the Improvement Notice indicate the following:

Failure to comply with the Notice will, under Section 33 of the Act, render possible a fine up to £1000 on summary conviction (i.e. when the case is heard in a magistrates' court), or an unlimited fine on indictment conviction (i.e. when the case is heard in a higher court)
An additional fine not exceeding £50 per day can be imposed where the offence is continued
The date of improvement to be achieved is at the Inspector's discretion; he can adjust the date when he feels it necessary
The recipient of an Improvement Notice is not exempt from liability for any part of his contravention
Conditions are outlined under which appeals can be lodged. An appeal suspends the actions of the Notice but does not affect the date of remedy imposed by the Inspector

Prohibition Notices (Figure 13)

Detailed references have been made (pages 17, 35). A Prohibition Notice stops work which contravenes the requirements of the Health and Safety at Work etc. Act. The Prohibition Notice is a legally binding document issued under the powers given to the Inspector under Section 19 of the Act. Supplementary notes in the Notice indicate the following:

Failure to comply with the Notice will lead up to a possible fine of £1000 on summary conviction and unlimited fine on indictment conviction
A continuation of the offence can involve an additional fine of up to £50 per day
The recipient of the Notice is not exempt from liability for any part of his contravention
Conditions are outlined under which appeals can be lodged
An appeal made does not suspend the Notice

Other relevant documents

The following documents are also a part of the statutory control of legal procedures:

Form 10 covers commencement of building operations (unless expected to be of less than six weeks' duration)
Form 41 covers (*a*) diseases such as lead and phosphorous poisoning, etc., and (*b*) diseases such as those caused by tar, oil, pitch, bitumen and paraffin-based products
Form 43 B covers (*a*) hoists and lifting appliances that are unsafe, collapsed or failed; (*b*) various situations involving explosions, regardless of injury or death; (*c*) certain causes of fires, regardless of injury or death; (*d*) immediate cause of death, or injury resulting in three or more days' absence; (*e*) work stoppages, exceeding five hours, for fires caused by celluloid, gas, vapours or where fire or explosion stops work, regardless of injury, for a period exceeding twenty-four hours
Form 2404 covers employment of young persons under the age of eighteen (see Figure 10, page 46).

6 Accidents–costs and reporting

An accident can be described as a situation or event which was unexpected, a chance, or an unintentional act. Much thought can be applied to each of these phrases, because each implies that accidental happenings just 'drop from the sky'! It is more truthful to describe an accident as a calamity that could, by following correct procedure and actions, have been avoided. There are exceptions of course, but it must not be forgotten that life, limb, working ability, money and more are lost because of 'accidents'. The construction industry records more than 30 000 accidents a year with still more thousands of minor, unrecorded accidents.

Legal requirements

The statutory control on construction employers is that an accident book (Form B1 510) is available at every workplace. Within this book all accidents, however minor are recorded. Regardless of how minor the injury may be, a record must be kept in case a more serious ailment develops, with the injury being a contributory cause. (See also Figure 5, page 39, for a typical accident book.)

Another legal requirement is that any of the types of accident listed below must be reported in the General Register (Form 36), and in Form 43B (typical examples shown on pages 52-3, Figures 14 and 15) which is forwarded to HM Factory Inspectorate. These are 'reportable accidents'; they are:

Accidents resulting in death

Accidents causing absence from normal work of more than three days

Accidents resulting in injuries which may not cause three days' absence but may create a reduction of wages or restriction of normal working

There are also items reported on the General Register (Form 36) and on Form 43B which should be reported to the Inspectorate. These are in part an accident report but can also be classified as an incident or situation report (page 59).

Company requirements

The requirements established by the company are influenced by the company safety policy, its safety standards and the size of company. Most organizations should have some form of accident review on record. A company should consider the following:

Personal injury – financial loss and reduction or loss of that person's skill

Plant or apparatus damage – restriction or loss of equipment

Insurance – future premium including financial or moral implications

Total loss – the calculated cost of the accident to employee, other employees and company losses

The detailed procedure varies in each company. Here are some general examples of the action which help in reducing future hazards.

Accident report

If an accident is considered dangerous, or serious, it is recorded on an accident report. (See Figures 16 and 17 for typical examples.) It is at the discretion of the site team leader as to what warrants such a report. The company may analyse the details of each, compute them, and produce a total company safety report on an annual basis. There are several advantages in such a scheme, which is operated by many safety-conscious organizations. (Figure 21, page 61)

PART 3

Accidents and dangerous occurrences

Address of operations or works on which accident happened (1)	Date of accident or occurrence (2)	Date of notice on F43B to HM Inspector of Factories (3)	Name of person injured (4)	Sex (5)	Age (6)	Usual employment (7)	Precise occupation at time of accident (8)	How accident or dangerous occurrence happened and what injured person was doing (9)	Nature of injury and whether fatal or not (10)

SPECIMEN

9

Figure 14 *General register form for reporting accidents and dangerous occurrences*

A notice in this form should be sent (immediately the accident or dangerous occurrence becomes reportable) to HM Inspector of Factories. (*See instructions overleaf.*)
NOTE:—If the accident is fatal HM Inspector should be informed immediately by telephone.

DEPARTMENT OF EMPLOYMENT

F 43B

FACTORIES ACT 1961, **section 80**
(as extended by SR & O 1947 No. 31)

Prescribed form of written notice of
ACCIDENT OR DANGEROUS OCCURRENCE
occurring in the carrying on of a Building Operation or Work of Engineering Construction

FOR OFFICIAL USE
District and date of receipt.

1 (a) Person (or company or firm) undertaking Building Operation or Work of Engineering Construction:
Name
Registered office or address
(b) Actual employer of injured person (if other than above):
Name
Address
(c) Trade of actual employer of injured person (*tick item which applies*):

Asphalt/tar sprayers	Electrical contractors	Plant hiring contractors
Builders (General)	Floor'ng contractors	Plasterers
Building and civil engineering contractors	Glaziers	Plumbers
Civil engineering contractors	Heating and ventilation contractors	Reinforced concrete contractors
Constructional engineers	Joiners and carpenters	Roofers
Demolition contractors	Painters and decorators	Scaffolding contractors
Other trade (*specify*)		

2 **SITE** where accident or dangerous occurrence happened:
(a) Address (and telephone number) of site
(b) Exact place on site

3 **NATURE OF WORK** carried on at:
(a) **Building Operations** (*tick items which apply*)
(i) Construction
(ii) Maintenance
(iii) Demolition
of
(iv) Industrial building
(v) Commercial or public building
(vi) Dwellings over 3 storeys
(vii) Dwellings of 3 storeys or less
(viii) Other
(b) **Work of Engineering Construction** (*specify type*)

SPECIMEN

4 **INJURED PERSON**
(a) Full name (*surname first*) **Sex** **Age**
(b) Address
(c) Occupation (*tick item which applies*):

Bricklayer	Carpenter/Joiner	Painter	Plasterer	Plumber	Scaffolder
Steel erector	Demolition worker	Steeplejack	Slater/Tiler/Other roofing worker		
Labourer (*specify trade where labourer worked for a tradesman*)					
Other occupation (*specify*)					

Note: *Semi-skilled men or apprentices should be classified under the appropriate occupation.*

5 **ACCIDENT or DANGEROUS OCCURRENCE**
(a) Date .... Time ....
(b) Full details of how the accident or dangerous occurrence happened. If a fall of person or materials, plant, etc., state height of fall. (*If necessary continue overleaf.*)
(c) State exactly what injured person was doing at the time.
(d) If machinery was involved, state:
(i) Name and type of machine concerned (inc. cranes)
(ii) Part of machine involved
(iii) Whether in motion by mechanical power at the time

6 **INJURIES AND DISABLEMENT**
(a) Nature and extent of injury (e.g. fracture of leg, laceration of arm, scalded foot, scratch on hand followed by sepsis).
(b) Was injured person disabled for more than three days from earning full wages at the work at which he was employed?
(c) Was the accident fatal?

7 **Has accident (or dangerous occurrence) been entered in the General Register?**
Signature of Contractor, Employer, or Agent **Date**

MR Group
Ref. to
M of T, etc.
1 Serial No.
2 MWBG
3 Age Group
4 F, NF, DO
4(a)
4(b)
5 Process
6 SIC
7(a) Causation
7(b)
7(c)
7(d)
7(e)
7(f)
7(g)
7(h)
7(j)
7(k)
7(l)
8 Occupation
9 Injury Nature | Site
10 Trade of employer
11
12
13

Figure 15 *Form for details of accidents and dangerous occurrences; which is forwarded to HM Factory Inspecorate*

R & D Building Company

Pararad Road, Roselip

SUPERVISORS ACCIDENT REPORT (FORM A) — Report No.

Date

Full name and address of injured person — Age

Occupation

Name and address of employer (if other than above)

State precise nature of injury — Fatal YES / NO

Where was treatment given and by whom

Date and time of accident

Location

Details of incident

Name and address of witnesses

If machinery, plant or equipment involved, complete FORM B (overleaf)

NOTE: Has accident been recorded in registers (tick appropriate box)

- ☐ Accident Book B1 510A
- ☐ General Register (Form 36)
- ☐ General Register (Form 43B)

Signature — Agent / General Foreman

Figure 16 *Supervisor's accident report form*

R & D Building Company

Pararad Road, Roselip

SUPERVISORS ACCIDENT REPORT (FORM B) Report No.

Date

Machinery, plant and equipment etc. involved in accident described on FORM A

Description of plant

Was it defective before use YES / NO

If YES state defects

Who was notified of defects and when

To whom does the plant belong

Brief description of subsequent damage

Figure 17 *Supervisor's accident report form – machinery involved*

R & D Building Company

Pararad Road, Roselip

SUPERVISORS INCIDENT REPORT (FORM C)

Report No.

Date

Date and time of incident

Location

Details of incident

Name and address of witnesses and / or persons involved

Cause of incident

Recommendations to avoid similar situation

Signature

Figure 18 *Supervisor's incident report form*

R & D Building Company

Pararad Road, Roselip

SUPERVISORS INCIDENT REPORT (FORM D)

Report No.

Date

Machinery, plant and equipment etc. involved in incident description on FORM C

Description of plant

To whom does the plant belong

Brief description of damage

Last recorded maintainence / repair

Approx. cost of work involved to repair

Recommendations to avoid similar incident

Signature

Figure 19 *Supervisor's incident report form – machinery involved*

R & D Building Company

Pararad Road, Roselip

SAFETY REPRESENTATIVES NOTIFICATION (FORM E)

This form is for the use of safety representatives or employees to notify the employer of unsafe and unhealthy conditions and working practices, and unsatisfactory arrangements for welfare at work.

Completed forms should be handed to your supervisor as confirmation of oral communication if immediate remedial action is required.

SITE DATE

GIVE DETAILS OF IMMINENT OR OBSERVED HAZARD OR MALPRACTICES

IF MACHINERY OR EQUIPMENT INVOLVED, WHO DOES IT BELONG TO

DATE OF INSPECTION SUPERVISORS SIGNATURE

SIGNATURE POSITION IN COMPANY

OCCUPATION DATE

Figure 20 *Safety representatives' notification form*

Incident report (Figures 18, 19 and 20)

A dangerous occurrence may take place without actually causing injury. The recording of the details and their causes on to a company incident report can lead to improved safety procedures and future elimination of such problems.

The costs of an accident

The cost of an accident to an employee

The *price* of an accident can be assessed in terms of replaceable losses. The *cost* of an accident cannot. Thus the cost to the employee can be reviewed in terms of *total cost*, including financial, physical, mental and other considerations:

Loss of earnings – the employee may have been responsible for the accident and may lose his normal wages

Loss of confidence and morale – the employee may have incurred an injury that will make him unsure and afraid of his future work

Physical disability

Loss of working ability – the employee may be unable to work normally again

Death – the ultimate cost

The cost of an accident to an employer

It is difficult to detail the real cost of an accident to the employer. A good employer has the welfare of his workforce at heart, but at the same time needs to consider the financial implications of an accident:

Lost wages – the employee may not be individually at fault; and he must legally or morally, during his absence from work, be paid the wages that he cannot earn because of his injury.

Wasted wages – other employees involved with the incident offer assistance to the injured man. Genuine help or even sympathetic observation cost the employer money during the work stoppage.

Supervision and administration loss – the organization and detailed arrangements for care, assistance and first aid to the victim take up supervisors' time.

Lowering of morale – any serious accidents are followed by reduced productivity through lowered morale, renewed caution and general concern for the injured person and his dependents.

Plant costs – if apparatus or machinery is involved in the accident, there can be delays in returning to normal working while: (*a*) an investigation by safety officers, inspectors or insurance personnel is made; (*b*) repairs are made to damaged equipment; (*c*) plant wrecked in the incident is replaced; and (*d*) a replacement operator is employed or trained to use undamaged plant (should the driver have been the accident victim).

Damage to materials – delivered materials, belonging to the client and paid for within an interim certificate, have to be replaced.

Reduced output for injured person – an injured victim returning to full employment and basic wage may initially need 'light work' or a similar low-geared output.

Insurance premium – may be increased as a result of the accident.

Incidental costs – reports, reviews or investigations, either at company level or with Safety Inspectors, are reflected in Head Office costs. Additional costs for first aid, telephone calls, form-filling, document-copying and adjustment of work procedures are incurred as a result of an accident.

The price of an accident

The cost of an accident includes all considerations in terms of machines, manpower, materials, time, space and money. It is the last factor which is the true price, because all considerations have a financial value. 'Money is the root of all evil,' is a well-known proverb but, amended, could read, 'The lack of money causes risk, causes death, and causes expense.' The short-cuts at work are the shortcomings of employees in terms of money. The following example is purely hypothetical but should nevertheless promote an awareness of actual price in monetary terms.

A simple 'low cost' accident may be as follows. A painter fell from a ladder that was not secured to the building. He received various injuries, including a broken wrist. As he fell paint was spilt over a large area of face brickwork.

The price of the accident to the employee is:

	£
Lost earnings for four weeks (allowing for supplementary payments)	130.00
Damaged clothing	20.00
Incidental costs (fares to hospital, prescription charges, etc.)	5.00
	155.00

The price of the accident to the employer is:

	£
Lost production – two painters helping for one hour (hourly rate includes all costs, overheads, etc.)	10.00
Transport to hospital – driver 2 hours	8.00
vehicle 15 miles at 20p	3.00
Repair or cleaning of brickwork	50.00
Incidental costs: first aid, materials loss, insurance, telephone	5.00
Allowances for wages to operative on reduced output on return to work (allow 2 weeks at £60.00 at ½ output)	60.00
	136.00

This simple example can be easily taken to any length and reviews known values. However, the remedy is not to value the accident, but to avoid it.

Accident investigation

An accident investigation is of little benefit to the immediate victims, but helps future prevention. There is always a reason for an incident, whether the responsibility lies with the operatives, management, or is the result of a lack of communication between them. Regardless of what has happened, a review of the supervisor's incident report should reveal facts that will assist any investigation. Additionally, it may be advantageous that the company prepares and uses a suitable investigation procedure. This could take the following form:

Review the supervisor's Incident Report
Question the site team leader or equivalent
Examine the injured person's normal work practice, experience, machine involvement, etc.
Question the victim and any witnesses
Review any malpractice
Review any malfunction of equipment
Review any combination of causes, whether verbal failure or human error
Report and recommend future avoidance

The following table illustrates levels of accident review and investigation. The standard and level of importance placed by different companies on these matters reflects their safety record.

Accident investigation	
Very good	Complete investigation with written report submitted within one working day and reviewed by management
Acceptable	Incidents involving injury or financial loss: written information for annual review
Unacceptable	No investigation
Accident reporting	
Very good	Accidents recorded or reported with follow-up to management and outside parties, e.g. (insurers, safety consultant)
Acceptable	Accidents recorded and reported as required by law
Unacceptable	Poor, inconsistent recording
Accident analysis/statistical review	
Very good	Clear reviews recorded and shown on graphs or similar pictorial chart for easy reference. Regular reports to management with indication

R & D Building Company

Pararad Road, Roselip

SAFETY COMMITTEE ANNUAL REPORT

Company Safety Officer:
Assistant Safety Officer:
Safety Consultant:
Company Doctor:

Summary of accidents as reported at monthly Safety Meetings

Number of recorded accidents	20
Number of reportable accidents treated on site (eg cuts, grazes, etc)	13
Accidents with hospital treatment but no other lost time	3

Description of injuries

Broken toe	Lost time 12 days	1
Broken ankle	Lost time 7 weeks	1
Head injury	Lost time 4 days (Sub-contractor)	1
Badly bruised hand	Lost time 7 days	1

Frequent bad practices noted

Scaffold defects
Trailing leads
Unguarded stairs and openings
Untidy sites

Conclusion

The health and safety of all employees and site visitors is of great importance, and employees as well as management should make every effort to enforce good practices at all times.

Works Director

Figure 21 *Company safety committee annual report*

	of trends and current performances which evaluate safety needs
Acceptable	Review and record of accidents with filing procedure, so that checking can be maintained
Unacceptable	No review except in serious loss of finance, work force, materials or where death occurs

Accident prevention schemes

Incentive schemes and competitions help managers to maintain safety standards. To create an awareness is success in itself, but to establish a relationship across the barrier that exists between management and workforce is just as desirable. Here is a list of ideas which could promote such a relationship. Such schemes already exist, either in part or in whole, in some construction companies.

Incentive schemes

Bonus payments for clean safety records create a keenness to avoid accidents. Payments to all operatives who avoid any incident over a six-month period can also be considered. These schemes have to be carefully controlled – they can tempt employees not to report accidents.

Competitions

A competition could be organized by the management between site team leaders for the best site safety practice. The winning site leader is left to choose whether he keeps the reward for himself or distributes it among his regular site personnel. There could be, alternatively, a non-monetary award, for example a night out for all site operatives paid for by the comapny. Any competition shoud lead to greater awareness of safe working practices. (See Figure 39, page 94.)

7 Health hazards

Hazards in the construction industry can be split into two main areas, 'health hazards' and 'physical hazards'. This chapter reviews 'health hazards', those hazardous situations in which internal damage to the employee can be caused, involving for example, disorder or malfunction of the lungs, stomach, ear or brain. 'Physical hazards' involving actual body structure are fully dealt with later and body protection reviewed (pages 96-101). This chapter deals exclusively with dangers to health and how the employee can be best protected.

Dust

Dust is created by the cutting, abrasion or dismemberment of a material, fabric or solid substance. Dust dispersed into the atmosphere is a direct pollution of the workplace and particles inhaled into the body may cause health hazards. Under the terms of the Health and Safety at Work etc. Act, the employer has a legal obligation to provide and maintain a safe place of work. Dust hazards can be reviewed in the following way:

Particle size

Dust particles exceeding 0.0051 mm in size are retained in the nostrils or the throat. Any smaller dust particles can pass through this natural defence barrier and enter the body tissues of lungs or stomach.

Intensity of dust

The various levels of dust intensity, discomfort and toleration, vary from individual to individual. The danger level also varies according to: each individual's strength of health; the type and source of dust; the type of work process; the type of work conditions; and the duration of exposure. HM Factory Inspectorate lays down acceptable limits of human exposure, using complex charts of data in which 'threshold limit values' (TLVs) are indicated for different hazards and acceptable periods of exposure. These are guidelines of the lowest acceptable limit of human exposure to a dust hazard.

Type of dust

Different dust particles react on the body in different ways. Lead in dust from rubbing down lead-based paints, for instance, is absorbed into the lung and subsequently circulated into the blood stream, causing poisoning. Silica dust from granite, asbestos and similar geological minerals has a long-term effect on the lung tissues. Some dusts do not cause danger to the body tissues, but to the skin (see page 67 for details).

Protection from dusts

The degree of protection varies according to the type of dust. The best form of protection against any hazardous dust is to abandon its use and find a substitute material. Crocidolite (blue asbestos) is so dangerous that it has been taken off the construction market almost entirely. To take this degree of preventive action against all harmful materials is not realistic or practical. Safety must be workable; although hazardous substances are necessary for construction work, there are safe working procedures for them.

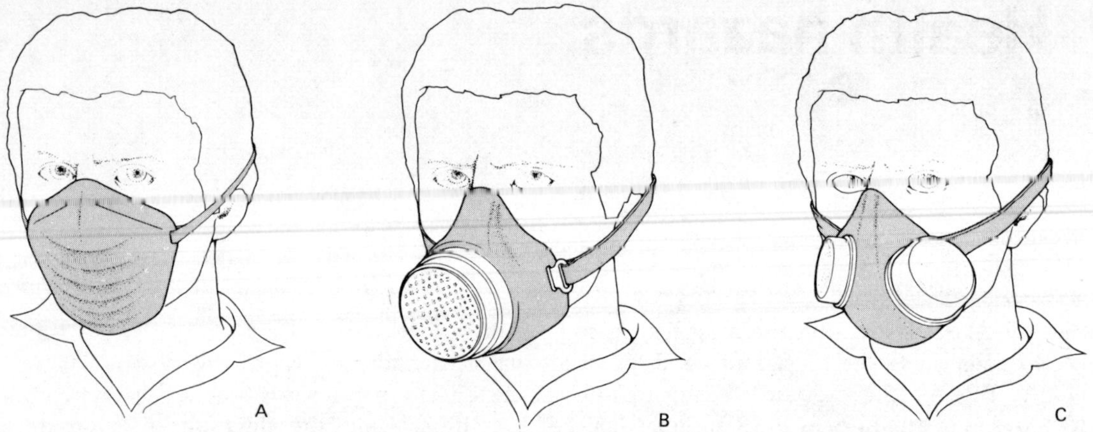

Figure 22 *Respirator equipment*

A – Simple 'throw-away' filter for non-toxic hazards and low-density dust contamination
B – Simple type of carbon filter strapped to the operator's face. The operator's nose and mouth are completely covered by this type, which has replaceable filters
C – Dual-type carbon respirator strapped over the operator's face in such a manner as to enclose nose and mouth
D – Complete hood arrangement which can be fitted with carbon respirators or to a separate air supply through air lines

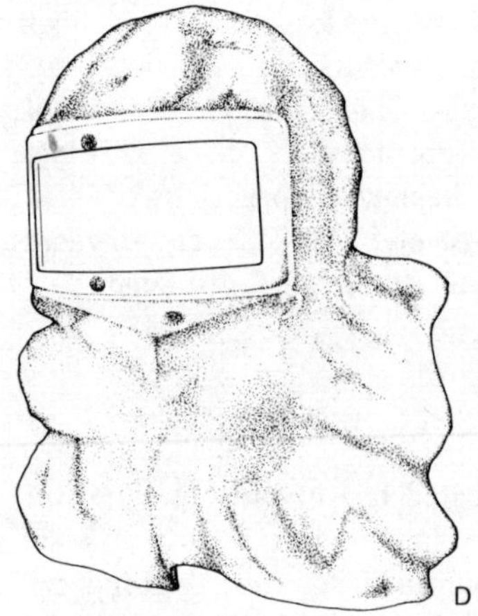

Asbestos

The use of asbestos and asbestos products is still extensive throughout the construction industry. Although crocidolite (blue asbestos) is no longer available, it is frequently found during repair, renovation or demolition of older buildings. This type of asbestos requires stringent working procedures.

Chrysotile and amosite (white asbestos) are commonly found in asbestos boarding, sprayed asbestos, asbestos rope and asbestos laggings. Asbestos-cement products, which contain approximately 10 per cent asbestos silica, are less dangerous, but certain precautions need to be taken. Special care is needed when working with asbestos or asbestos-cement products which may have been sprayed, dyed or coloured during manufacture, thus creating a confusion with pure crocidolite, which is a rich dark blue colour.

Handling asbestos

Pipes, sheet materials, moulded forms etc., can be handled safely as long as care is taken with personal hygiene before food breaks. Protective gloves must be available to all employees. Fibrous asbestos products, e.g. rope, yarn or quilt, must not be handled without the following precautions, or equipment:

A good filter respirator, issued personally to the operative

Strong protective clothing which is sealed around the wrists and ankles. After use the clothing

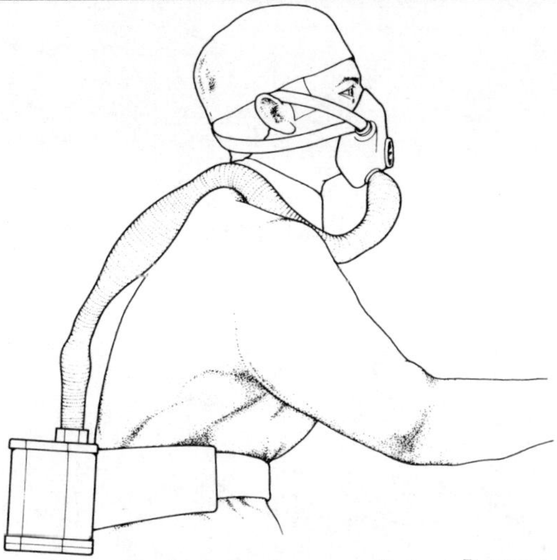

Figure 23 *Positive-pressure respirator. Battery-powered air supply provides a low pressure to reduce the possibility of inward leakage of dust, etc. round the mouthpiece*

must be sealed in impermeable bags, marked 'asbestos-contaminated clothing', and cleaned under special arrangement
Protective gloves must be worn at all times
The hazardous material must be stored in sealed bags if in small quantities

Hand-cutting asbestos

Provided that the employee protects his hands with a suitable barrier cream or wears gloves, the hand-cutting of asbestos is acceptable as long as:

It is performed outside or in a well-ventilated room
The workpiece is pre-dampened
Offcuts and scrap are stored in polythene bags for disposal at an authorized tip

Machine-cutting asbestos

Small amounts of cutting with powered hand-tools can be performed without hazard, as long as:

The work is pre-dampened with clear water, to reduce the dust movement
The cutting apparatus is fitted with an exhaust ventilation device to retain all dust particles created by cutting
Suitable protective clothing is worn

Larger amounts of cutting or continuous cutting require the same powered hand-tools. The precautions, however, are more involved and include:

A separate dustproof shed, or booth fitted with an exhaust ventilation system as near as possible to the workpiece
Protective clothing, as previously described
A respirator, or breathing device in cases where the exhaust system retains the contamination level described in the Asbestos Regulations 1969 (Figures 22 and 23)
All waste and scrap asbestos must be stored in polythene bags, sealed and disposed of at an authorized tip

Choosing the correct respirator

The recommended limits of air contamination for different types of respirators are shown in the table below.

Types of respirator	*Maximum no. of fibres per millilitre of air*	
	crocidolite (blue)	*chrysolite or amosite (white)*
Half-mask dust respirator	4 (unless demolition)	40
Positive-pressure powered dust respirator	20	200
High-efficiency dust respirator	80	800
Self-contained breathing apparatus	more than 80	more than 800
Compressed airline breathing apparatus	more than 80	more than 800
Fresh-air intake hose	more than 80	more than 800

Drilling asbestos

Should asbestos need to be drilled, similar precautions to those for small amounts of cutting activities are required. The drill must be fitted with a special shroud or hood with an exhaust-ventilating unit, and/or the operative needs to wear a suitable dust mask or respirator.

Stripping asbestos

When demolition or stripping-out is in progress, the most likely location of asbestos is in boilers or lagging. If crocidolite (blue asbestos) is to be removed, HM Inspectors must be informed twenty-eight days before work begins. The following checklist should be observed for the removal of asbestos of all kinds:

Samples must be taken and analysed so that the degree of protection needed can be established
The operatives must be briefed of any hazard and issued with necessary safety apparatus and garments
The work area must be cleared, and any obstructions or unmovable equipment must be protected and covered
If the asbestos is to be removed from part of the work area, screens must be erected to enclose the asbestos-contaminated area
Adequate warning notices must be erected, e.g. 'Danger – asbestos' or 'No entry unless protective garments worn'
Asbestos must be soaked before being removed
Asbestos debris must be collected into sealed polythene bags, and labelled for authorized disposal
All dust from floors and ledges of windows etc., must be stored away from personal clothing
At the completion of asbestos stripping or demolition, protective clothing must be vacuum-cleaned and sent in sealed, labelled polythene bags for specialist laundry cleaning

Masonry, stone, brickwork and concrete

Work causing dust hazards includes: hand operations; cutting or chasing by powered hand-tool; and external cleaning. The dust created is less harmful than asbestos dust, but precautions should nevertheless be taken.

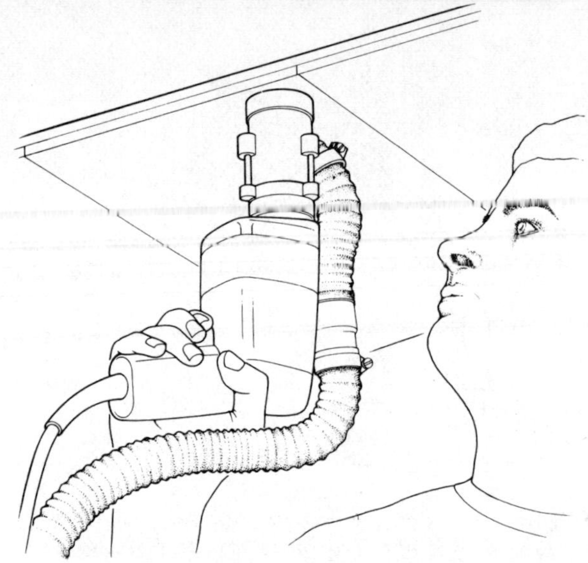

Figure 24 *Preventing dangerous emissions into the air. A vacuum attachment to drill or cutting machine sucks away dangerous dusts and contaminants*

Hand cutting and chasing

Protection of eyes and provision of good tools are covered later (pages 70, 98) but dust control can be achieved by:

Pre-dampening the work area
In excessively windy conditions or in a confined area the wearing of a mouth mask incorporating gauze padding

Machine cutting

Machine cutting is carried out by powered hand-tools with abrasive wheels or high-speed cutters. Clipper-saws do not create dust because they have constant water spray devices (page 186). The following precautions are needed:

Protective clothing, goggles, mouth masks, and in certain cases gloves, must be worn
The area must be well ventilated (but the dust must not be blown away into another work area)

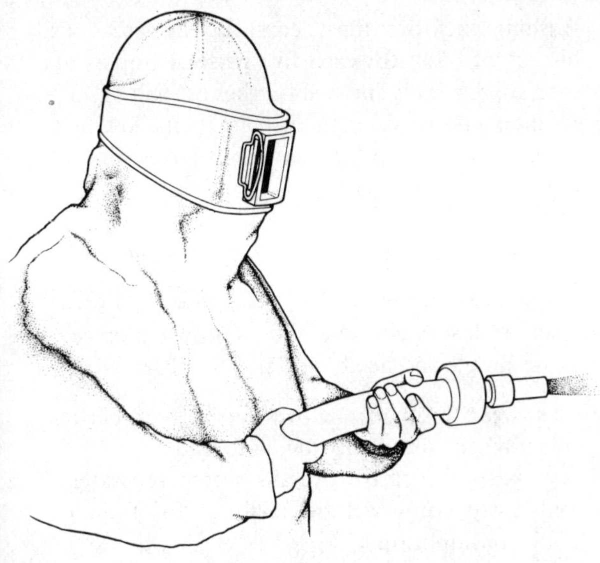

Figure 25 *Sandblasting requires this full protection. Another vital safety precaution is to use non-silicous grits*

In confined areas an extractor must be used to remove dust particles

Grit blasting

Grit blasting is a specialist means of cleaning the external faces of existing buildings. It can be hazardous and requires extreme caution:

Grits containing free silica or any form of silicous grit must not be used

A complete facial respirator with clean air supply must be worn (this type of respirator also protects the operator's eyes and head) (Figure 25)

All persons in the vicinity must at least wear suitable eye and preferably wear complete face protection

Impregnated timber and certain hardwoods

Certain hardwoods when machined can create dusts which are harmful if inhaled. All timbers that have been treated with preservative are dangerous to the body. Skin diseases and irritations (page 100) can also be caused by treated timbers. Chemicals used to destroy wood-boring insects and wood-destroying fungi are also poisonous to humans. Treated timber dust particles, if inhaled into the stomach or through the lungs into the bloodstream, can cause poisoning. Some liquids which form into crystals on the wood surface also crystallize on bare skin.

Hand cutting

Handling and hand working of impregnated timbers is, in general, relatively harmless provided that the following basic precautions are taken:

Wash hands immediately after working with treated timbers, and always before any food break

Wear protective gloves if the timber is being treated with a brush, and if 'wet' timber is used

In addition, where spray treatment is being performed, mouth mask protection is needed

These points mainly refer to hand protection, but there is an internal health hazard by poisoning or stomach disorder

Machine cutting

The detailed requirements for woodwork machines are dealt with later (pages 179-83). The following protection is necessary, in conjunction with general wood-machine and powered hand-tool safety:

Where machines, powered tools, sanders, etc., create dust a suitable extractor scheme must be used, as close as possible to the source of dust, or on the machine

All operatives must wear mouth masks, if it is not possible to extract the dust

There must be good ventilation

Additional hazardous dusts

The following dust hazards require similar precautions to those already mentioned

Dust from plastics

Plastic fillers and particles are harmful to the lungs. A mouth mask must be used when sanding, filing or machine-cutting most plastics.

Dust from resin

Resin-based fillers are not to be inhaled. Mouth masks must be worn, when sanding these fillers; in confined areas it may be necessary to wear respirators with dust cartridge infill.

Dust from glass fibres

Fibrous particles of glass fibre, rock wood and similar insulation materials can cause throat irritation and possible lung infection. Mouth masks should always be worn and hands *rinsed* clean to wash away all infectious particles. Excessive rubbing whilst washing can cause skin irritations.

Lead poisoning

Contamination to the body can be caused either by dust from materials that are protected by lead-based paints, or by direct contact with lead and lead-based materials. Intake of dust particles into the lungs and into the bloodstream is by direct breathing. Intake of lead by direct contact results in contaminated hands coming into contact with the mouth (Toxic fumes from lead are dealt with below.) The main group of construction workers involved with lead are painters and plumbers (pages 194, 201).

Lead poisoning by contaminated dust

The health hazard created by dust contaminated with lead results from dry rubbing of painted surfaces. Modern paints contain less lead than those of past years, but the hazard still exists and the following precautions should be observed:

A respirator, with self-contained breathing apparatus, must be used where lead-contaminated dust circulates close to the operative
Lead paint must be stored in clearly marked tins or containers
Protective clothes must be worn and regularly washed
Protective clothing must be stored separately, well away from welfare facilities for food and drink consumption
Hands must be washed before all food breaks. Washing facilities must comprise at least one bucket or basin for each five persons employed, with soap, towels and nail brushes at each basin
Dry rubbing-down of modern paints with low lead content can be done with the protection of filter-type mouth masks

Lead poisoning by hand to mouth contamination

Any operative using lead or lead-based materials is open to lead poisoning. The following precautions are simple and need to be strictly observed:

Always wear protective gloves or a skin barrier cream when continually handling lead
Always keep protective overalls stored separately, well away from welfare facilities for food or drink consumption
Always wash hands before taking any food break. Washing facilities must include at least one bucket or basin for each five persons, with soap, towels and nail brushes at each basin

Toxic fumes

Several operations create toxic fumes. There are numerous combinations of type, severity and level of poison which affect the duration of hazard and degree of cure. The plumbing and services engineering trades are most vulnerable to this hazard (pages 201, 209).

Toxic fumes from welding, flame cutting and lead burning

The following checklist is a general guide in preventing toxic fume hazards:

If the work is in a confined area or a workshop situation the fumes must be extracted
If, in such confined situations, a suitable exhaust system is not possible or practicable, the operative must wear a respirator incorporating an air line or self-contained breathing apparatus
Smoking must be forbidden. Contamination can be caused by inhaling the fumes via the cigarette
Clothing must be stored separately, well away from any welfare facilities
Hands must always be washed before food breaks

Noise

Noise can be termed unwanted sound or a confused sound. Sound is a sensation produced by rapid fluctuations of air pressure, the vibration of which causes the ear to react to its frequencies. Sound is recorded in decibels (symbols dB) which is a logarithmic scale. This ranges from zero, which is the threshold of audibility or commencement of hearing, up to 130 dB, which is the level of pain. Construction activities create noise which, depending on extent, duration, pitch, frequency, or intensity has varying effects. Tiredness, nausea, and headaches are common, short-term side effects of unacceptable noise levels. The long-term effect can be temporary or permanent deafness.

Reducing noise

Ear protection is necessary in several situations and advisable in others. The following recommendations are effective individually or in combination.

Contain the noise producer

Encasing the machine, apparatus or equipment within a confined noise booth reduces the level of noise. Employees in the vicinity may be able to work without ear protection. Airborne noise, which vibrates the air immediately surrounding the noise source, is contained within the booth. A noise booth needs to be of cavity construction, using separate leaves with a thick infill layer, or lining, or insulating quilt or sheet insulation. The access doors should be tight-fitting and sealed during the running period of the machine. Alternatively, the doors can be double layered internally and externally. Materials to be fed into the machine, e.g. wood machines for six cutting, can pass through a hatch with flexible panels. This reduces the escape of noise as the workpiece enters and leaves the noise booth (Figure 26).

Isolate the machine from the floor structure

Sounds passing through the structure fabric, known as structure-borne noise, can create discomforting vibrations. Structure borne noise also vibrates surrounding air, subsequently causing air borne noise hazards. Heavy static machines anchored directly to the floor structure, cause structure vibrations. The machine base needs to be isolated from the main floor structure. A surround layer of cork or similar fibrous sheet material absorbs noise between machine base and working floor structure (Figure 29, page 73).

Separate 'noise' room

Some machines are too large or impracticable to contain within a noise booth. In such circumstances it may be acceptable to make provision for

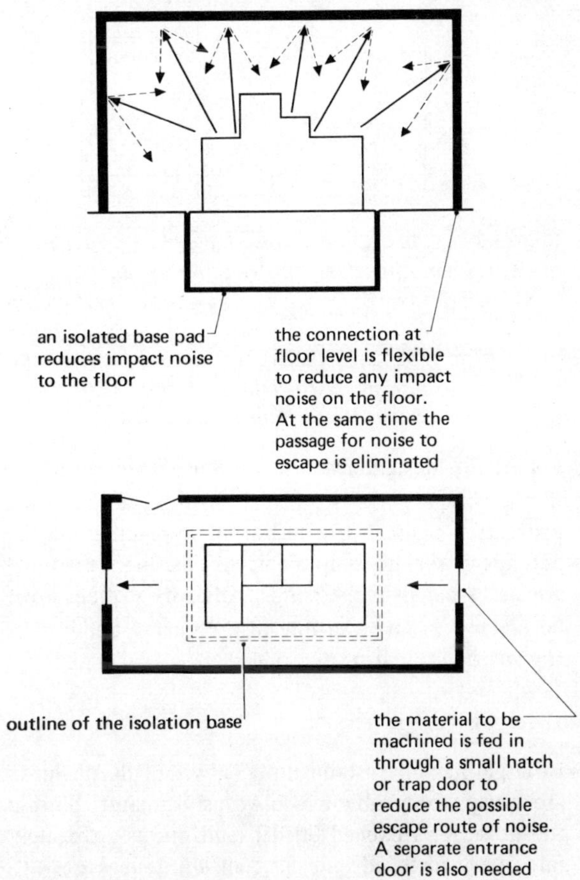

Figure 26 *A noise booth. It is constructed of lightweight materials which contain the noise and are flexible enough to contain vibration. A typical method of construction is to build a lightweight timber shell, which is then lined with heavy layers of fibreglass or rock-wool quilt*

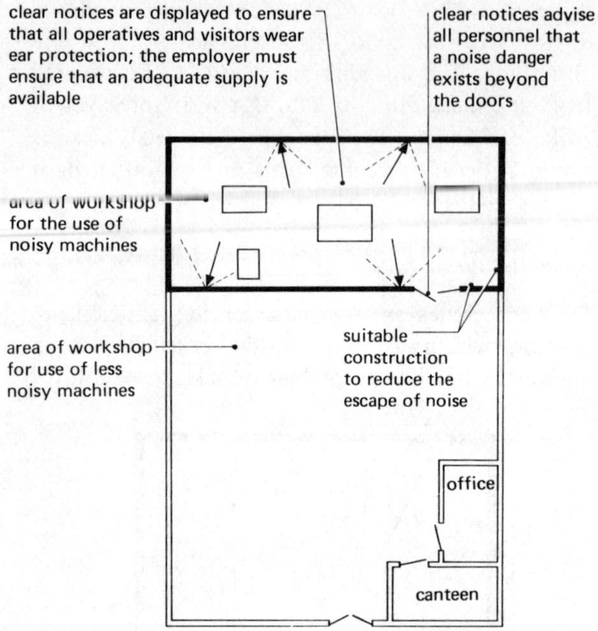

Figure 27 *Controlling a noise hazard by isolating the noisy machines. The 'noisy zone' is built of acoustic brickwork. Its windows are of acoustic construction with two frames separated by a clear gap of 0.210 m. The doors between the two parts of the workshop are designed to reduce noise-penetration; self-closing devices keep them tightly shut.*

a specially designed room, or noise zone to house them. Employees, however, still need personal protection against the airborne noise created. Ear defenders for all employees or visiting personnel are necessary in such an area. Advisory notices must be erected to give warning of the hazard and ensure the wearing of ear protectors.

Total protection

In large machine establishments, where all machines are noisy, the problem is obviously greater. Should the noise level exceed 90 dB continuously throughout the normal working day, all employees need to be issued personally with suitable ear protectors. The wearing of such equipment must be made compulsory by the company safety policy and terms of employment. 'Total protection' refers to the complete area requiring ear defenders for prolonged periods.

Spasmodic protection

Certain machines which create excessive noise may only be in use for short, intermittent periods during the working day. The employer in such cases should make available a good method of ear protection while the noise exists. For the most part only the machine operator is vulnerable to excessive noise. Clear notices must be erected instructing the use of ear protection, and a disciplinary procedure to deal with offenders set up. The machine manufacturer has a legal obligation, under the Health and Safety at Work etc. Act, to give full details of noise levels created by his equipment and advise how the levels can be reduced.

Vibration of powered hand-tools

Several powered tools produce considerable vibration. The operator, taking a firm stance to use the equipment, resists and himself absorbs the vibrations which are being set up. The operator's nervous system is put under excessive pressure during the working period, and this can result in his hands losing their 'feel' or the sensation of touch, the condition commonly known as 'dead hand' or 'white hand'. General weakness caused may lead to an accident as a result of the operator's temporary incapacity.

Whilst such equipment is used the operatives on this type of work are vulnerable to risk. Certain actions can be taken to reduce the in-use hazard of powered hand-tools. These are as follows:

Heavy glove protection

If it is safe and practical to do so, the operative should wear heavy industrial gloves, which reduce the vibration. This practice does not eliminate the problem entirely but it makes the work process more comfortable and reduces the risk.

Intermittent use

A work procedure must be established which allows substantial rest periods, possibly one which entails two operators working intermittently on the vibrating equipment. The period of individual actual work contact with the hazard is thus reduced.

Bad weather

The results of being exposed to inclement weather may be regarded as a body problem although internal organs are also affected. Those employed in construction expect to get wet during outside work activities. What is not appreciated however by the same people is that this hazard is a contributory factor to many deaths and is one which can be avoided. The discomfort is great and it is remarkable that site operatives still expose such risk to the body. The Construction (Health and Welfare) Regulations require that adequate protective hutting for shelter be provided during bad weather but situations do exist where work must nevertheless continue. In these situations protective clothing must be provided. Failure to protect the body may result in direct illness of colds, flu and general short-term discomforts. Continued lack of protection may develop into rheumatism, bronchitis, arthritis, with the prospect of future complications.

The simple maxim for this area of health risk is 'prevention is better than cure' – either get under cover or wear the right protective clothing. The following points are for general guidance during bad weather conditions. They rely very much on common sense and an understanding of each other's problems by both employer and employee.

Wear protective clothing

The Construction (Health and Welfare) Regulations require that protective clothing should be issued to those working in rain, snow, sleet or hail. Unfortunately the design and restrictive nature of many weatherproof clothes tempt operatives to discard them. Lightweight equipment making possible unrestricted movement is available, and management should discuss with site team leaders, safety officers, and operatives the most suitable pro tective clothing before arranging purchase.

Provision of drying rooms

Adequate facilities for the drying of personal clothing or protective clothing must be provided. Should this not be practicable, the employer is obliged to provide some other means of drying clothes. It is of mutual advantage to employer and employee and in the interest of good, continuous working if suitable drying facilities are provided.

Provision of protective screens

In those cases where work must, for any reason, continue despite weather conditions, a protective screen may be necessary. A screen not only protects employees but also materials, equipment and finished work. Conditions on large areas of scaffold, in exposed situations, or open sides of partially completed buildings, may be improved by protection from polythene or similar sheet material.

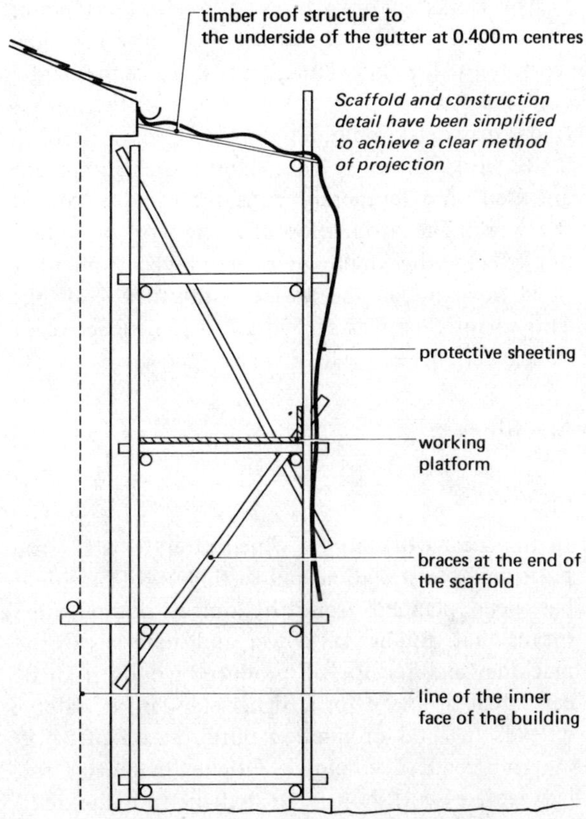

Figure 28 *Protective screening on a scaffold. Where the type of work allows it, the scaffold can be used to form a weatherproof screen. A lightweight timber support carries heavy polythene sheeting or tarpaulins across from a secure fixture under the gutter and down the outside of the scaffold*

8 Workshop dangers

The resources available to management that allow production, manufacture or fabrication of goods, can be classed under the following broad headings: machines; manpower; materials; money; space; time.

The workshop procedures covered in this chapter are off-site activities, carried out in established workshops for individual trades. In some larger companies, a complex of workshops for the various trades may exist. A building contractor may establish a site workshop in which apparatus and equipment are used on a temporary basis for the duration of the work. The installation of a site workshop does not exempt the employer or employee from their obligations under the various statutory controls. This chapter applies to workshops irrespective of their location, use, style or trade activity.

Machines

Installation

In the ideal workshop – which rarely exists – the positioning, installation and setting-up of machines has been planned from the outset. Pre-planning means that supplies of power and materials to the machine, and receipt of produced goods from it, can all be arranged for a higher standard of safety; it leads to good production output, ease of working, and reduced employee fatigue. In most cases, however, a workshop is created by sporadic additions on an *ad hoc* basis. In any case the following factors apply to all machines of a static fixture.

Secure fixing

All machines need to be anchored to a firm subfloor, with adequate bolting around the base of the machine. Several fixing schemes are available. The most common is a projecting threaded stud from the floor over which the machine is positioned and held in place by a double-nut device. Where vibration is expected, the manufacturers supply spring-loaded mountings as shock absorbers between machine and floor. If excessive vibration is expected, the machine needs to be mounted upon an isolated base, separated from the surrounding floor by a lining of cork or similar, shock-absorbing material (pages 69 and 73).

Power supply

Power must be supplied to the machine through an independent supply of electricity or compressed air which has been connected by a trained fitter. All supply pipes or cables must be kept clear of possible damage, and fitted with a protective sheathing or plate in the most vulnerable places. Supplies are best provided through floor ducts with removable top covers for easy maintenance access. Overhead supplies are acceptable but may need protection from moving parts of other machines or any materials. Where machines are wall-mounted or adjacent to a wall, the supply is best fed down to the machine with a secure fixture to the wall or in a service duct within the wall structure.

Maintenance

After installation a machine needs to be checked and set up. Manufacturers' literature provides information relating to its good running and safe working. The future of the machine and the safety of those who use it depends on good maintenance procedure. There are two approaches to maintenance:

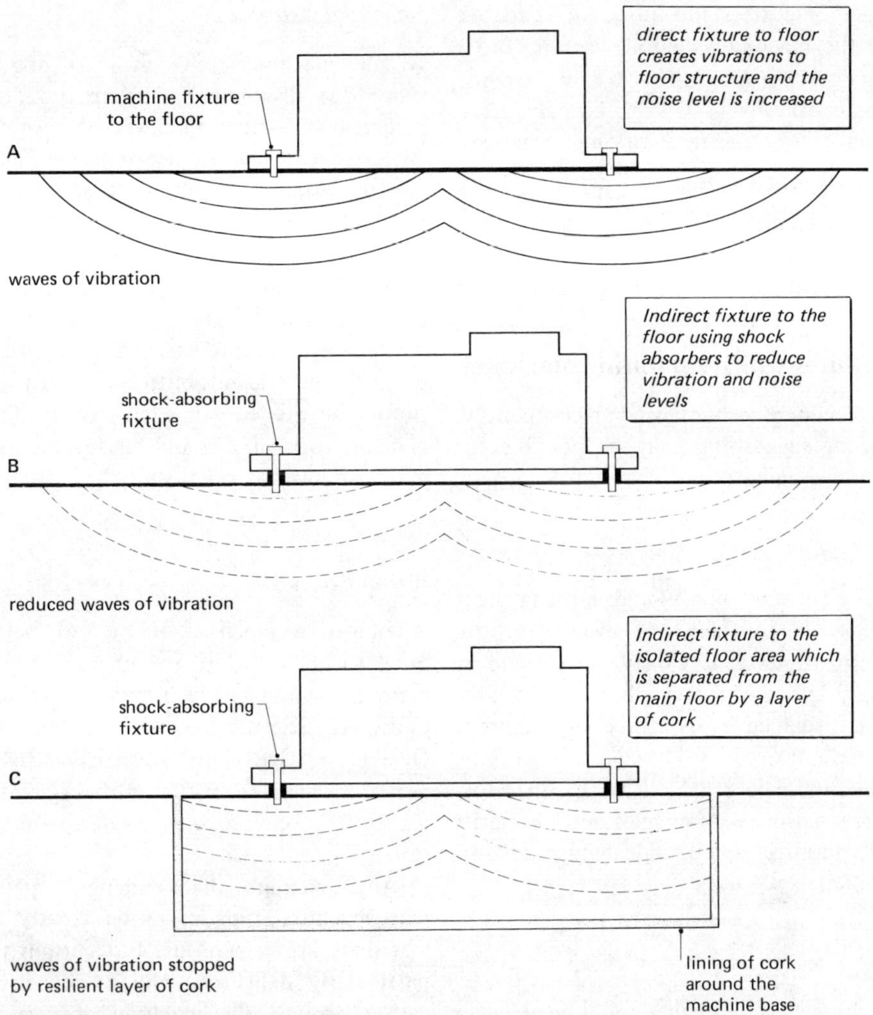

Figure 29 *Fixing machines to workshop floors.*
A – Direct fixture. Vibrations travel immediately to the floor structure. Noise level is high
B – Indirect fixture. Shock-absorbing mountings reduce vibration and noise
C – Indirect fixture to an isolated floor area. The resilient layers of cork prevent vibration spreading

Planned maintenance is a prepared plan of maintenance activity for all machines in an organized procedure with regular checking at the intervals recommended by manufacturers.

Preventive maintenance is maintenance work administered to any machine when action is needed to avoid a breakdown or malfunction threatening safety. Preventive maintenance also applies to cutters and moving parts that need sharpening as they wear.

Planned maintenance is by far the best scheme and is recommended for the following important reasons:

Preplanned maintenance avoids a breakdown of any moving part and should reduce risk to the employee.

Maintenance records, from planned work, provide a concise record of machine workload, repairs, replacements, requirements and discontinuation of obsolete parts or machines.

Planned changes of cutters and lubrications reduce risks, keep the machine in good working order and prolong its working life. A safe machine provides the operator with a safer job, free from hazard, injury or unsafe working practices.

Preventive maintenance is an accepted scheme of maintaining machines but is less organized than planned maintenance. There are, however, situations in which it has advantages.

Safety procedures whilst operating machines

It is difficult to eliminate human error completely, but the following suggestions may be possible for certain types of machine.

No power until safe

By integrating electrical supply and moving parts it may be possible to retard or even stop the machine movement, until all safety devices are in position. If any part of the machine e.g. guard, fence etc., is not aligned to its safety requirement, the machine does not operate. These devices do not, however, eliminate the personal risk to the operator, who should observe the use of goggles, mouth masks or any aids like push sticks and guides while actually using the machine.

Flashing lights

A series of flashing coloured lights could be electrically connected to the machine. These would warn the operator of any hazard and would switch off when all safety requirements had been observed and implemented on the machine.

Audible warning

Operating in a similar way to the lights, or in conjunction with them, a sound could be transmitted, until the safety of the machine is acceptable. When all safety requirements were observed, the sound would be automatically turned off. Such sounds would obviously have to be at a suitable level of audibility to warn the offender, but not distract other operators.

Emergency stops

A machine must have at least one independent switch to disconnect the supply in the event of emergency. There can never be too many of these switches for good safety practice. The main power can be turned off by a wall-mounted switch or a pendant switch hanging near the machine. All emergency switches must be clearly marked with notices advising what they are for and how to operate them. It is good practice to test each switch on a regular basis. This can be achieved by using the emergency buttons to turn off the power supply at the end of each day or shift. A simple rota of 'turn offs' could be devised to ensure each stop switch was tested regularly.

Warning notices

Whenever a machine is out of action through breakdown, maintenance check or simple adjustment it is good safety practice to let all employees know. A clear notice, such as 'DO NOT USE – OUT OF ORDER' or 'DANGER – BREAKDOWN' must be visible from the normal operating position, leaving the operators in no doubt about the situation.

Any machine that endangers persons in the close vicinity must also be clearly marked. All machines are hazardous, but certain processes are particularly dangerous and a clear area of demarcation around the machine or apparatus should be made. Spraying and welding machines are particular examples of 'long-range' hazards requiring warning notices (pages 192, 205-6).

Artificial lighting

Workshop lighting in general is dealt with on page 80. Certain machines also require individual lights for setting up positions. It is extremely dangerous to attempt setting up and test runs with inadequate lighting. Direct lighting at particular points around a machine can be achieved by radial arm lighting or a lead light moving about the parts of the machine or equipment.

Environment control – extractors, etc.

Employers have an obligation not to subject employees to a hazardous environment. Certain machine processes emit welding fumes, cellulose spray fumes, timber and organic dusts, etc. into the work area. Pollution of the work space must be controlled by the extraction of all harmful dusts or gases. This is achieved by vacuum extraction at a point as near as possible to the source of such health hazards. Detailed schemes are reviewed later.

Manpower

Adequate supervision

To maintain a good level of safety practice all workshops must be supervised and have a leader. This person, whether fully or partially in a supervisory role, is responsible for all those under his direction. The status and powers of leadership vary in different situations or organizations. These responsibilities are similar to those of a foreman (page 32). The accepted ratio is one supervisor to between five and nine subordinates, enabling the supervisor comfortably to maintain leadership over his subordinates, depending on workload, type of work and style of workshop.

Adequate training

Young persons must be prepared and trained to operate machines: it is not acceptable for anyone to work on a machine process without adequate instruction; those under eighteen years of age are only allowed to operate most machine appliances when under personal supervision or training. Once a machinist has completed training, and is over eighteen, he should be proficient and competent at his work activity. College training can prepare any machinist for such work and give theoretical instruction for examination purposes. In-service training and instruction are also important in developing ability and knowledge of safe work procedures.

Mature operators also need training and guidance in machine working. Machinists of long experience may be reluctant to accept training, but this is essential for their future well-being. Familiarity breeds a contempt which, combined with a complacent attitude, can cause hazards and dangerous situations. Refresher courses at colleges and national training schemes have an important role to play. A good safety practice is to organize the machine workforce into a scheme which ensures periodical training or refresher courses at regular intervals.

Controlled age limits

Legislation controls at what age a young person can commence working a machine. No legal control, however, limits employment at a machine potentially hazardous to older operatives. Unfortunately reactions can deteriorate with age. It could be prudent therefore to arrange for older or mature machinists to use less hazardous machinery. This should in no way degrade the experienced skill of any machinist and must not devalue the individual in status, security or financial grade. It is not possible to set an age limit upon safe working of any particular operation or operator. The management should however maintain a careful review of those employed on machines at all ages.

Materials

Storage of large commodities

The various trades and work activities in construction depend upon different material commodities, each having a separate procedure for storage. Each trade is reviewed in later chapters. This section deals with the general concept of materials storage in workshops.

Racking

The materials used must not obstruct the working area of a machine or production workshop. Wherever possible large items – timber, sheet materials, metal tubing, metal sheeting, etc. – must be kept in a racking suitably made to retain all parts of the material without collapse.

Pallets

Certain materials – e.g. bricks or boxed packages – can be stored on pallets. These are timber (or sometimes metal) flat tables with blocks underneath to allow a forklift truck to insert its prongs. With a forklift truck the pallets can be lifted from the floor and moved around the workshop. The working area should be planned to allow clear moving access for these pallets (pages 78-9). Palletized transport can also be used for movement or distribution of completed goods from the workshop, or partly completed components about a workshop complex.

Trolleys

Wheeled trolleys incorporating a flat table with lockable castors are ideal for transporting materials or partly completed components. The floor needs to be flat or reasonably level to allow safe movement. Certain types of trolley have spring-loaded wheels which reduce the chances of the load being dislodged by vibration. The better-quality trolleys are fitted with hydraulic supporting frames which keep the trolleys consistently at the most suitable working height and reduce employee fatigue (see also pages 149-51).

Storage of small commodities

There are hundreds of small parts, units, fixtures and materials that require a good stores arrangement for easy access and safe procedures. The stores will be established with racking or shelves according to the details of the workshops to be serviced. Access to the racking or shelves must cause no hazard. A suitable arrangement is shown in Figure 65 (page 122), in which a rack ladder for access to all shelves is secured by a hook to the top of a common rail running along the line of storage shelves. Care is needed to ensure that incompatible goods are separated. This is particularly important with certain liquids and gases.

Issue of hazardous items

Certain materials are dangerous in themselves and special storage arrangements are important.

Register of hazardous goods

There should be a register of the entire store contents of each material known to be dangerous. A typical page of a store issue book is shown in Figure 30. This needs to be kept up to date with details of quantity held, amounts issued and recipients. Several adhesives, solvents, paints, etc., require this degree of care. It may be advantageous to incorporate the storage of certain dangerous tools, e.g. cartridge ballistic gun, within this scheme. In the event of misuse, or loss of dangerous articles from the work area, the issue book confirms who was responsible for them at the time.

Advisory notices

Should any materials stored require protective garments, e.g. goggles or a special glove for hand protection, it is good safety practice to have a clear notice to instruct the storeman. A simple directive, 'Rubber gloves must be used with this chemical' or 'Goggles must be worn when using this', will remind those concerned of safe working activities.

Warning notices

Hazardous materials must be clearly marked with suitable notices. Asbestos and similar goods requiring careful handling should have a notice, 'Caution – wear gloves' or 'Caution – wash hands after handling'. Gases and similar goods that may easily ignite or explode must be clearly marked, 'Danger – Explosive goods' or 'LPG – NO Naked flames'.

Finance

The safety of construction employees is not always directly related to monetary values. It is however, indirectly, a very important aspect of the safety, welfare and well-being of construction personnel, both on site and within workshops. Financial control is a management function. This needs good communication between section leaders who need the money to spend, and the management team who control expenditure. The following factors are related to safety.

R & D Building Company

Pararad Road, Roselip

STOCKS OF DANGEROUS MATERIALS

Date	Issue from Stores (out)	Receipt to Stores (in)	Balance	Issued to	Signature of receiver

DANGEROUS MATERIALS ISSUE LIST

Issue No.	Date	Goods Issued	Issued to	Signature of receipt	Return of goods (if any)

Figure 30 *Log of stock and issue of dangerous materials*

Adequate finance for facilities and maintenance

There should always be funds available to establish the full facilities and needs of an individual workshop or a group of workshops. Capital expenditure should be organized by management to incorporate the safe establishment and upkeep of workshops. There should be no skimping, and the full issue of safety needs should be available. There should also be suitable finance for maintaining the best safety standards.

Finance for training

As well as training facilities, there must be suitable financial support for the instruction required. The safety officer or training officer, who co-ordinates such courses of instruction, is responsible for all organizational and financial support. If too little money is available, the training of employees may be restricted and subsequently substandard.

Financial incentives

There should be some provision to reward employees who maintain good safety procedures.

An *accident-free bonus* may be paid to any workshop employee who avoids an accident or dangerous occurrence for a predetermined period. This system must not, however, discourage employees from completing relevant forms or reporting any dangerous situations or accidents.

A *prevention bonus* may be paid to any employee who conscientiously observes the safety procedures required at his place of work. This needs a system of checking by the workshop supervisor.

Space

Space availability varies. The space at sites is determined by the area the client makes available. Workshop space, however, is provided by the employer, who can establish a workshop of suitable size and standards. Other than the obvious monetary restrictions, the following limitations may exist:

Size and clear floor span of the building
Amount and size of machinery or apparatus to install
Amount of clear working space needed around these machines
Areas needed to manouevre materials or partly completed work
Limitations of material storage or completed work
Type of work to be undertaken and goods to be produced

The safety implications related to workshop space can be reviewed through the following points.

Layout

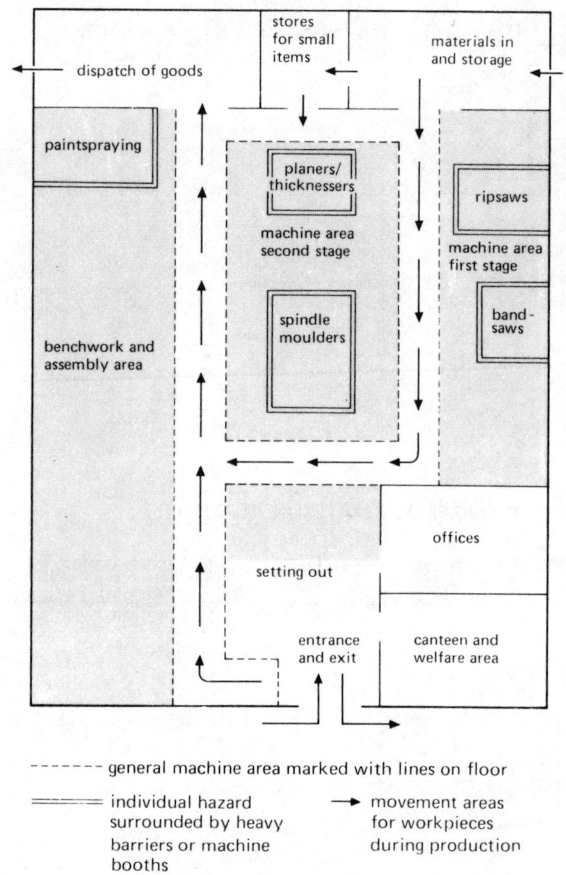

Figure 31 *Safe and efficient workshop layout, in this case for joinery work. General areas of work process are logically arranged. Individual hazardous machines and activities are isolated by heavy barriers or machine booths. White lines demarcate the lanes for material movement and machine-prohibited work areas*

The lay-out of any workshop or work area is determined by the activities carried out in relation to the size of goods to be produced and the components involved in manufacture. The need for progressive movement through the channel of production affects the lay-out of machines so as to attain an even flow through the workshop. Restrictions must not be allowed to hinder the activities, and there must be no deviation from safe working. The working area of machines, men and moving of equipment or materials must be made safe for the job to be done.

Corridors for movement

It is important to establish clear areas of movement around machines and similar work places. Corridors alongside machinery, booths or workbays must be clearly marked with white or yellow lining on the floor. Barriers must be erected around dangerous zones, preventing transport access. Allowance must be made so that the swinging around or traversing of materials or goods avoids collisions.

Marked hazard areas

Certain machines must only be operated by accredited personnel. No one else should be permitted near the active machine. In such cases, a clear zone must be marked and reserved for the operator. All other employees must be kept well clear. This is also necessary around 'distance hazards', e.g. welding, where glare may be dangerous. Clear areas of demarcation must be made in accordance with either the machine manufacturer's literature, or the discretion of a competent person, or of the safety officer responsible. In extreme cases of danger it is necessary to isolate in a separate room all high-risk equipment and to allow admission only to those working with it.

Fire escape

The preceding section stresses the importance of good areas of working and movement. It is very important nevertheless to retain a clear escape route. In the event of fire all operatives must have a clear, unrestricted exit, unimpeded by machines, materials or completed goods. The lay-out of any workshop must make adequate provision for easy escape. Every employee must have a clear directive about what action to take in an emergency. Fire points should be established in the workshop and an adequate provision of fire-fighting appliances. The type of fire extinguisher necessary for each workshop varies. Clear notices of instruction must be shown at each fire point, for each type of extinguisher to be used (see also pages 164-5).

Adequate notices

Simply worded notices of instruction can provide a great deal of advice and information. These notices fall into the following categories: advisory; instructional; command.

A colour code for each style of notice aids recognition and understanding, e.g. red notices for command (a strong, dominating colour), blue for instructions, yellow for advice.

Non-slip floor

In several work activities a slippery floor and consequent unsteadiness of the employees' feet creates a hazard. The initial floor covering must be specified in accordance with good work procedures and after consultation with the safety officer. If the floor becomes slippery or the use of the workshop is changed, floor coverings with abrasive particles can be laid or painted on. At the operating position of certain machines, special floor matting can be laid to give strong non-slip resistance.

Environment

Employers must provide a suitable environment for all employees and must not allow health hazards to develop from poor working conditions. The broad definition of environment refers to temperature, light, ventilation and noise, each of which needs to be reviewed separately.

Temperature

Different trades require certain standards of temperature for safe working. Some processes create warmth and prevent the employee from getting too cold. If the employee becomes uncomfortable or distracted by his discomfort he may fail to observe safety requirements. The workplace should therefore be maintained at a comfortable working temperature. The Woodworking Machine Regulations require that a minimum working area temperature should not fall below 13°C, and most other modes of employment need similar minimum temperatures. Where the work activities give off heat, mechanical fans should be used to bring in cooler air (mechanical inlet) or take out the hot air (mechanical extract). In general the workshop temperature should be moderate, not fluctuating excessively or causing distraction or discomfort to the employee at work.

Light

Natural light must be provided as much as possible to give unrestricted vision for much of the working day but in a way which reduces direct sunlight glare or discomfort from excessive solar rays. North light roofing is the most suitable but tinted glass can be introduced to any areas of windows that may be offensive.

Where artificial lighting is used, certain hazards may be created. One hazard is caused by the 'stroboscopic' effect. When the frequency of the current powering fluorescent lights coincides with the revolution of a machine part, the part may appear to be stationary or only to move very slowly – in much the same was as the spokes of wheels in a cinema film appear to move slowly or not at all. The stroboscopic effect can therefore lull workers into thinking fast-moving machinery is almost stationary. The danger is avoided by wiring fluorescent lights in a 'lead lag' system or by incorporating conventional filament lights (which are less prone to the effect) in the areas around the machine. Artificial light must be clear and adequate for all areas.

Where local lighting is needed flexible arm lights can be used individually as required.

Ventilation

Unwanted or hazardous smells or particles from the air must be extracted to maintain a clear working environment. The workshop process determines the intensity of air pollution, and, in turn, the extraction needed. Various schemes exist. A system with mechanical extraction is preferable to one of natural extraction with a mechanical intake, for the control of air purification. A system of fans acting as mechanical extractors are necessary in most workshops. In adverse conditions this extraction may be supplemented by a mechanical fan inlet which pulls in fresh air. The workshop air conditions should not cause any health hazard and need to provide employees with an acceptable working environment, regardless of the impurities created by work activities.

Noise

Several types of work develop noise above the acceptable limits of human comfort or which may cause a health hazard. This has been reviewed in detail (pages 69-70).

Welfare facilities

Clean, hygienic rooms for food breaks or medical needs must be provided, and also clean toilet and washroom facilities. These are statutory requirements, similar to those needed on-site (page 175). These facilities are long-term and should be established with a degree of permanency in an area adjacent to the workshops. They must conform to the legislation of the Health and Safety at Work etc. Act, the Construction (Health and Welfare) Act and/or the Offices, Shops and Railway Premises Act.

Time

Time can be valued in a number of different ways – by financial measures; progress; time saved or lost; and several other considerations. The correct safety view is not how much time costs, but how much can be saved in terms of human suffering, loss or disability. Time taken to do a job correctly establishes a safe work procedure. Other uses of time to the best mutual advantage of employer and employee are as follows:

Time for maintenance

Irrespective of what scheme of maintenance is implemented, adequate time allowance must be given to ensure that a thorough, complete job is done. If the time for maintenance is reduced, the person fulfilling the work will be pressured into an error or poor work. It is unwise of any management to attempt short cuts in maintenance. A job done well can be rewarding in terms of good production, as well as in the safety of those using the workshop.

Time for training

Statutory controls exist which prevent personnel who are under-age or lacking in experience and competence from using a workshop machine or apparatus. There should be no short cuts to training. A syllabus of full training requirements must be completed before anyone can be confirmed competent and able to use machinery or equipment needing trained operators. The initial training period must fulfill all the detailed requirements that make the operative safe at his place of work. A follow-up of interim periods of training, on a refresher basis, should also be allowed. There should be no time restrictions on fulfilling training standards.

Time for cleaning

All areas of the workshops must be clean, and particular attention must be paid to hazardous areas around machines or apparatus. In addition to the workshop areas the management are legally and morally bound to keep clean supplementary areas, e.g. canteen, rest room, stores. A provision of cleaning staff or time allowance for those employed in the workshop must be arranged for the well-being of employees.

Time for planning

Well-planned workshop lay-outs produce better spacing of equipment and materials incorporating good safety procedures. Adequate time for work-flow, planning of training, planned maintenance, schedules of machine replacement must be allowed. Hurried planning of these activities can cause dangerous and unsafe situations for those using workshops.

9 Site dangers

Construction management divides the resources available for the building process into six main components: machines; manpower; materials; money; space; time. This chapter reviews the dangers that exist on site and discusses under each specific heading how subsequent hazards can be eliminated. Safety is a serious topic needing a sensible approach but – notwithstanding all the legal jargon – the construction industry must be allowed to retain its workability. The employee must be allowed to work without undue restriction created by endless controls. Safety must be observed and respected, but above all be workable.

Machines

Electricity

Although electricity is not a machine, it is the chief source of energy for apparatus and machines.

Application should be formally made to the local headquarters of the Electricity Board for supply. Company management decides if this supply is to serve temporarily or to be part of the completed permanent installation. The provision of power to the site is completed by the Electricity Board, which terminates its work at a position planned by the contractor. This incoming supply unit must be protected from weather and physical damage; it houses the meter, fuse box and distribution panel. From this point the contractor, or his electrical subcontractor, establishes a distribution scheme to provide power around the site without hazard or danger. The accepted procedure for safe site working is a reduced voltage scheme in which all power is transformed down to a lower voltage. The apparatus needed to perform this must be provided by the main contractor (employer) and is best achieved by double wound transformers. Figure 33 illustrates how best to deal with the distribution on a typical large contract.

Woodworking machinery

The term 'woodworking machine' is defined by the Woodworking Machines Regulations 1974 as any machine (including a portable machine) of a kind in its schedule 1 (reproduced as Figure 32) used on any one or more of the following: wood, cork, fibre board, and material composed partly of any of those materials.

SCHEDULE 1

To be read in conjunction with Regulation 2(2)

Machines which are woodworking for the purpose of these Regulations:

1 Any sawing machine designed to be fitted with one or more circular blades
2 Grooving machines
3 Any sawing machine designed to be fitted with a blade in the form of a continuous band or strip
4 Chain sawing machines
5 Mortising machines
6 Planing machines
7 Vertical spindle moulding machines (including high speed routing machines)
8 Multi-cutter moulding machines, having two or more cutter spindles
9 Tenoning machines
10 Trenching machines
11 Automatic and semi-automatic lathes
12 Boring machines

Figure 32 *Extract from the Woodworking Machines Regulations 1974*

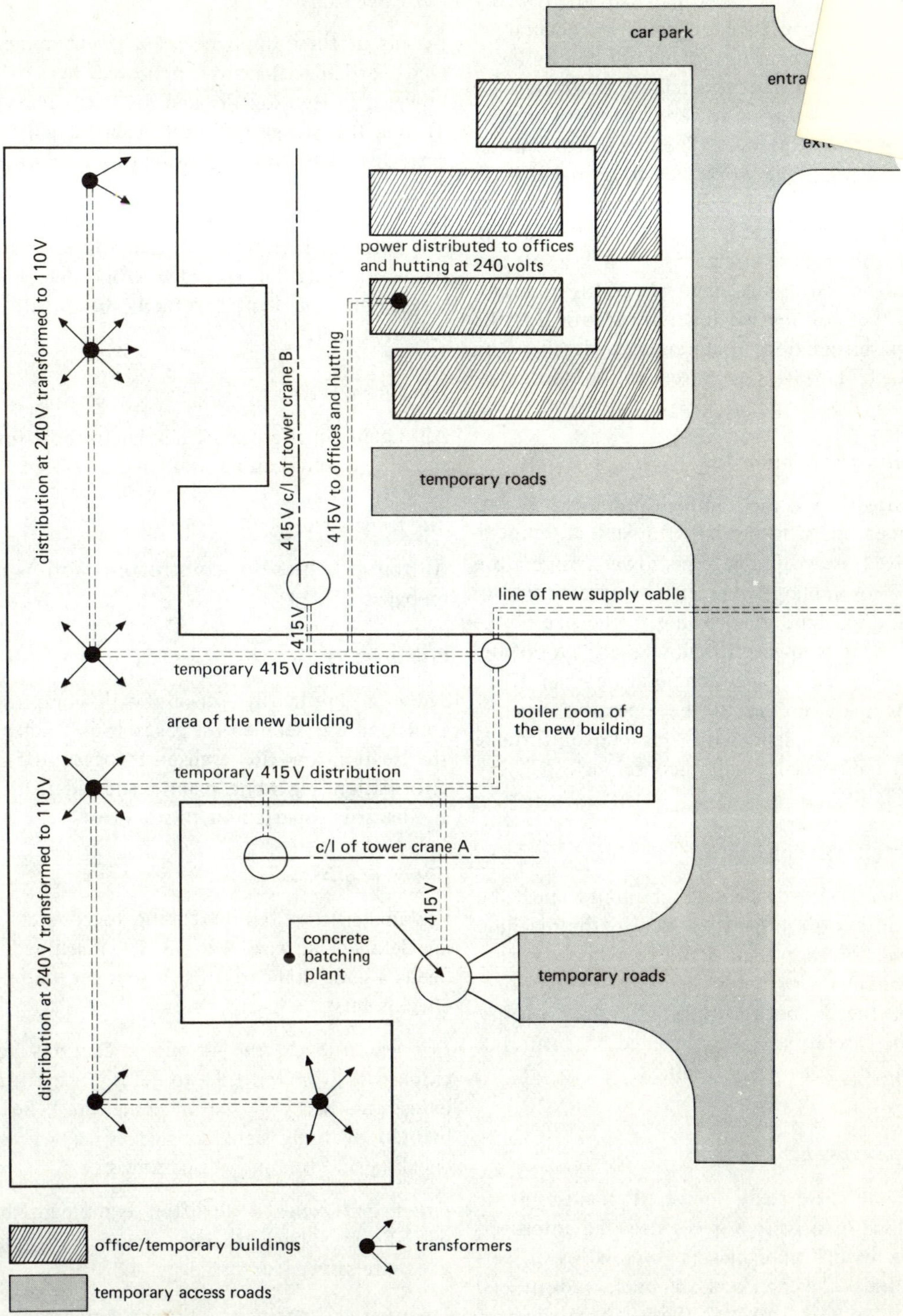

Figure 33 *Electrical supply on a large contract site*

'Powered tools' are those which can be physically handled by the operative, whereas a 'machine' is a larger piece of apparatus that cannot be physically lifted. Another way to understand the difference is that, in most cases a powered hand-tool passes over or along the workpiece. The machine, however, is static and needs the workpiece to be fed over or between the cutters of the machine.

The two most typical machines found on a construction site are a circular saw and a planing machine. The restrictions, controls and regulations, regarding these or any machine, are identical to the workshop installation requirements detailed on pages 71-81, 177-84. Certain exceptions do exist:

Temperature (Regulation 12)

The required workshop temperature is 13°C, but where 'open air' working is permissible a temperature of 10°C is acceptable. This lower temperature applies to log milling, where it is impracticable to enclose the machine area, owing to the size of the tree trunks. It is unwise to allow a site 'workshop' to be worked at a lower temperature than 13°C. A low temperature leaves the operative uncomfortable, preoccupied with his uncomfortable condition and hence less safety-conscious.

Extractors (Regulation 41)

Suitable methods of clearing chippings must be provided in a workshop. If, however, the machine for planing is other than a thicknessing machine, or is a circular saw, there may be exemptions, because neither is specifically listed in the statutory document. Discretion is then needed to ensure all practical precautions are maintained.

Portable powered hand-tools

The use of electrically powered hand-tools is increasing as contractors recognize the improved output from such appliances as circular saws, percussion drills and breakers, sanders, and others. The safety requirements for these are detailed on pages 134-41.

Cartridge tools

Details of these appliances are given on page 142. One particular hazard applicable to site use is, however, often overlooked by site management. This is the storage of both ballistic gun and the cartridges. Each must be stored in a separate place. All unused cartridges must be placed in a cool dry place which is locked, and well away from the gun. The issue or distribution of gun and cartridge must be controlled and recorded. Any defective cartridges must be kept separately for return to the manufacturer.

Abrasive wheels

All details on site use are similar to those for workshop use covered on pages 45 and 187-8.

Site transport

All transportation for construction work is covered on pages 125-33.

Compressors

These are primarily for use with compressed air tools and are detailed on pages 146-8. Refer also to the section on site transport (pages 125-33) for information regarding maintenance, defect control and hazards appertaining to site plant.

Concrete mixers

Certain parts of safety relating to concrete-mixing are detailed on pages 125-33. The machine however needs a good standard of safety under the following checkpoints:

Hoppers/drums/skips – all precautions to stop employees from working too close or even under a dangerous part of the machine must be implemented. Warning signs are needed as well as visual checking and training of operatives.

Pulley and ropes – all lifting equipment must be periodically checked and recorded to maintain adequate safety (see also pages 42-3).

Location and position – the machine needs a good standing and blocked wheels to avoid overturn or collapse. An additional consideration is the safe access to, and egress from, the machine of transport supplying and being supplied by the concrete mixer.

Working areas – the aggregate bays and certain areas around the machine need to be fenced, to prevent operatives being crushed by, or caught up in, any moving part or material.

Operatives awareness – all users of the machine need to be conversant with its dangers and aware of potential hazards.

Hoists

Additional scaffolding is needed to accommodate the hoist by way of a tower (page 109). The machine element is dealt with in the section on site transport (pages 125-33). Reference to these sections gives a complete appraisal of hoists as they apply to general site dangers.

Bar-benders/croppers and fabricators

These machines can be electrically or hydraulically powered. Regardless of its source of energy, the motor needs to be accurately and fully protected to avoid entanglement of the operator or other employees. Additional guarding may be necessary to the actual cutter, benders or shaping forms. Also the complete work area needs to be fenced off. The materials used must be in good condition. Serious cranks or bends created during transit need to be straightened before use to avoid serious lashing of the moving steel during bending.

Manpower

The resources of manpower fluctuate on the site according to the activities of the contract. Different trades arrive and leave site according to the work schedule. Many of the requirements and safety standards are developed under separate trade headings in following chapters. However, certain safety concepts are common to all trades and subcontract employments involved. The following points are therefore essential for *all* employees. They are based upon safety requirements and are not intended to be guidance for work procedures.

Supervision

Within the legal requirements of the Health and Safety at Work etc. Act the employer is legally bound to secure adequate supervision for all employees. There does not appear any legal control over what is considered an adequate degree of supervision. A rule for guidance is that one apprenticed trainee should be trained with not more than four craftsmen of his trade. Another consideration is the management concept that each supervisor should have between five and seven subordinates. This ratio varies according to the level of decision making and depth of responsiblity. For trade operatives, a maximum of ten should be controlled by one section leader.

Training

Under the same statutory powers, adequate training must be provided (see pages 21-2). Training establishments should incorporate safety in their lessons but the employer may increase the training by the following arrangements.

Induction courses – immediate training at the start of the trainee's or new employee's period of employment. This would include the issue of safety equipment and full understanding of the company safety policy.

Periodical courses – arranged, organized and usually given by the safety officer or his selected speaker, as a refresher on safety thoughts and ideas.

Policy courses – management seminars to update, improve or reiterate the policies of the company and enhance the understanding throughout management structure.

Discipline

A procedure of disciplinary control, to encourage good safety activities and to discourage malpractices, must be implemented. There are several methods which the contractor can employ to enforce a control, but all have limitations.

Monetary fine or incentive – it is almost impossible to deduct money from the wages of an employee for disciplinary reasons. There is however no reason why an additional payment for good safety procedures cannot be implemented, e.g. bonus for wearing helmet, bonus for a full year free of accidents, to individuals.

Suspension if the employee is contractually bound by an agreement, it may be feasible to suspend him from work without pay.

Termination of employment – the ultimate deterrent is for an employee to be dismissed as 'unsuitable for employment because he is a dangerous worker'. Several actions must be taken by the employer to ensure he completes termination correctly. An employee is entitled to a verbal warning, followed by a written warning, before dismissal.

Notify the Inspectorate – under the strict wording of the Act, an employer could, where he feels necessary, report an employee for failing to meet the legal safety requirements. This is hard justice, but necessary if the employee blatantly refuses to respect a demand from the employer. It is feasible for the Safety Inspector to issue, personally, an Improvement Notice to an employee.

Materials

The use of materials differs from trade to trade, so a detailed review is given in each trade chapter. However, certain aspects of material planning, delivery and storage are common to all trades:

Planning delivery

Delivery of material to the site should be preplanned. This enables both a safe work procedure and a suitable pool of labour and plant to be organized. Prepared delivery schedules are not only good management practice, but also essential to safe working. If there is no pre-planning, the materials may be off-loaded in an unsafe way. This probably puts the operatives at immediate risk and causes the materials to be stored badly, creating a hazard for future operatives. Adequate lifting gear or operatives must be available before off-loading commences.

Storage

A high percentage of fatal accidents is caused by falling materials. There are numerous examples of poor storage procedures in each trade.

Distribution about site

The quality of material distribution is a contributory cause to untidy sites. By careful thought, or simple calculation, the right material can be distributed to the right part of the site in the right quantity. Excessive materials are wasted and create a work hazard.

Finance

The need for safety improvement has created a financial burden on all employers to impose the correct procedures enforced by law. Unfortunately, there are still those employers, whom the law cannot control, who place monetary value and profits above human risk. Hasty construction work is an attempt to save time, and in turn save money. However the risk-taking during this rush can cost lives, on which no financial value can be placed. All construction management teams do allow for provision of adequate safety needs at tender or estimate stages. Most construction managers do plan for all practicable and legal requirements protecting the work force. Many contractors, however, fail to keep abreast of their legal and moral obligations for financial reasons. Many employees are forced to work dangerously and also to break the law, by exposing risk to themselves and others at work. It is illegal to take risks at work and equally so for any employer to force an employee to take such risks.

To sum up:

The employer must provide sufficient money to enable all facilities to be available.

Provision must be made at tender stage for good working conditions.

Failure to make the provisions legally prescribed could result in substantial fines.

It is illegal for any employee to take any risk at work. He should take all reasonably practicable precautions. No financial incentive or gain should be substituted for safety standards.

There are legal controls which protect the employee from the cost of safety provisions required under the Act. The financial burden should be borne by the employer.

Space

The area made available for work on site is subject to agreement in site lay-out drawings, and as part of the contract. The following factors relate to the safe use and application of space resources.

Existing services

These have a controlling influence how the site area should be used in relation to damage and hazards. All existing services – electricity, water, telephones and so on – must be located and clearly marked where they cross the site either beneath the ground or overhead or in existing buildings. Where possible, for safety or necessity, the supply should be terminated at the site boundary or re-routed around the site perimeter. Damage to service pipes is costly enough itself but also can create serious hazards, particularly in the case of electricity and gas supplies.

Telephones must be re-routed around the site before any operations, causing damage or danger, commence. Water supplies must be re-routed or if not used, properly sealed at the site boundary by the local water authority engineer.

Gas services must be diverted or sealed off. All pipes left within the site boundary must be made harmless, in case naked flames are used during work.

Sewage pipework must be diverted or sealed off at the nearest or most practicable inspection chamber. Any large excavations, interceptors or cesspools in the old system must be clearly fenced, or filled in, to avoid an accident later.

Electricity cables above ground level may be acceptable if they do not interfere with the building work during or after the contract. Any live cable within the site boundary must be clearly indicated with temporary masts and headboards or 'flag' indicators (see pages 132-3).

Electricity cables beneath ground need to be terminated at the site boundary, rendered 'dead' or re-routed around the site. All cables left in the site working area must be checked and confirmed 'dead' before any excavation is allowed.

Where it is impracticable to re-route existing services completely around the site, it must be agreed between the client, architect and service supplier where services can cross the site. Solicitors may also be involved in this operation to deal with wayleave, licence or permission.

There is a legal obligation to record all routes of services in the agreements of each service supplier. It is good practice also to record on site lay-out drawings, or within the site diary, exactly what action is being taken.

New services

There has always been a difference of opinion among site management personnel about the optimum timing for installing or laying new services. If services are installed early, excavation activities can be completed, covered and allowed to settle. If services are laid late, the site ground is disrupted, excavators have to be recalled and settlement continues after the contract is completed. This means further maintenance to landscape, roadways and footpaths. However, services laid early can be easily damaged by future earthworks, landscape work or heavy vehicles. Whichever decision is taken, it is important that all services are protected and clearly marked to avoid damage and safety hazard. All services can be marked at the time of installation with clear notices, at ground level, along the line of the cable, pipe or duct. It may be helpful to mark the line of the service with the British Standard colour code for building services:

Yellow – gas
Orange – electricity
White – telephone
Green – water

Hutting

Provision of hutting is a difficult problem, which can cause limitations and safety restrictions.

The main contractor's legal obligation is to provide space sufficient for a subcontractor to have reasonable hutting. The actual provision of huts is left to each subcontractor for his personnel and storage. The Construction (Health and Welfare) Regulations 1966 make the main contractor responsible for washing facilities, mess room, shelter, toilets, etc. If the burden of financial, legal and moral restrictions becomes unacceptable, major subcontractors may have to share responsibility.

Generally it can be assumed however that the main contractor will provide the following site facilities:

Washing facilities

On sites employing operatives for more than four hours, a washing facility must exist. If the work is of a maintenance or 'jobbing' nature this can take the form of water containers on the vehicle. If the work is expected to last for six weeks or more, or if twenty persons are employed, the facilities must at least include basins, or similar water containers, soap, towels or dryers, with a supply of hot and cold water. If the work is expected to last for one year or if 100 persons are employed, the facilities must include four basins, soap, towels, or dryers, with hot and cold water. The figure of four basins (one per twenty-five persons) becomes one per thirty-five if the work force exceeds 100: 50-75 persons need three basins, 76-100 need four, 100-135 need five and 135-170 need six. If workers use lead or similar poisonous substances, one basin or similar water container for five operatives, with soap, towels or dryers, hot and cold water and nailbrushes, must be provided. All facilities provided must be maintained in good clean condition.

First aid room

A contractor employing more than forty operatives out of a total site workforce exceeding 250 employees must provide a first aid room. This must be controlled and supervised by a qualified person (page 175). Construction office workers should be included as part of the site workforce when calculating first aid needs. The detailed requirements for a first aid room are included in the Construction (Health and Welfare) Regulations 1966 and are detailed on page 175.

Ambulances

A contractor employing more than twenty-five operatives should give written information to the local ambulance service offering site address, type of work and probable completion date. This should establish a good communication-link with the health service. The contractor may also appoint a man responsible for ambulance calls and other communications. Also to be remembered is that when more than twenty-five persons are employed, at least one stretcher must be provided for use in an emergency.

Toilets

The facilities for acceptable sanitary standards are similar to those for washing amenities. A suitable sanitary convenience must be provided for each twenty-five persons. Once the total site employment exceeds 100, a suitable sanitary convenience is required per thirty-five persons: 50-75 need three toilet fittings, 76-100 need four and 100-135 need five. All conveniences must be suitably screened, private, have adequate lighting, be ventilated and clean. There should be similar separate provisions for female toilets. All facilities must be free of any cost to the employees.

Shelters

On any site there must be suitable shelter for protection against bad weather, protection for personal clothing, storage of protective clothing and enough room for seating at meal-times with a good supply of drinking water. Wherever the contractor employs more than five men some means of heating the shelters and drying clothes must be provided. Where five or less men are employed these facilities must be observed as far as practicable. If the site employment by the contractor is more than ten, some means of heating food must be provided, unless hot food is available on site.

Tidiness on site

The expression, 'A tidy site is a safe site!' is relevant to all aspects of safety at work. Care and time taken on tidying up is well spent. Materials, plant, tools or machines left in an untidy state can create hazards.

Work spaces

Hazardous places of work require some form of temporary support or protection. The space provided for employees must be prepared, maintained and inspected by the employer or his delegated site leader. The access to, and egress from, the work space is equally important. Under the Health and Safety at Work etc. Act the employer must maintain the place of work and access to that position. Various categories and types of work space need special attention.

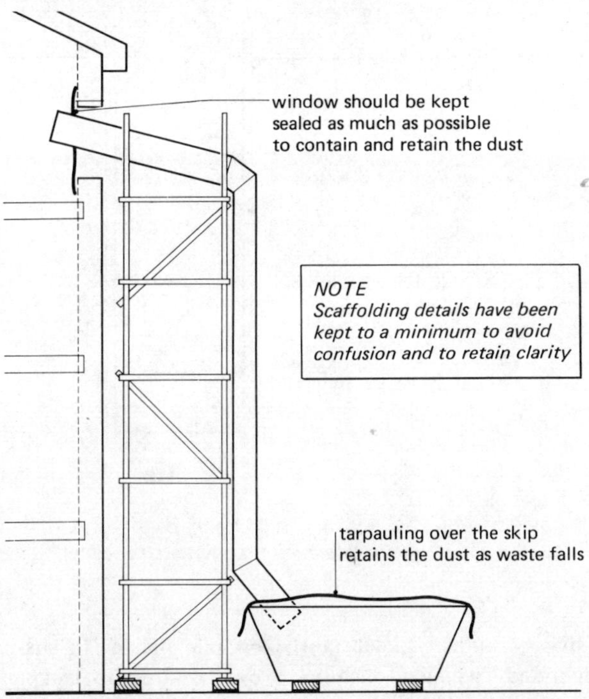

Figure 34 *Temporary chute for waste. The top of the chute is set into a window opening; the bottom is removeable so the skip can be changed when full. A chute is a quick and safe method of lowering waste, and reduces dust levels. (To keep the drawing clear, scaffolding details are omitted)*

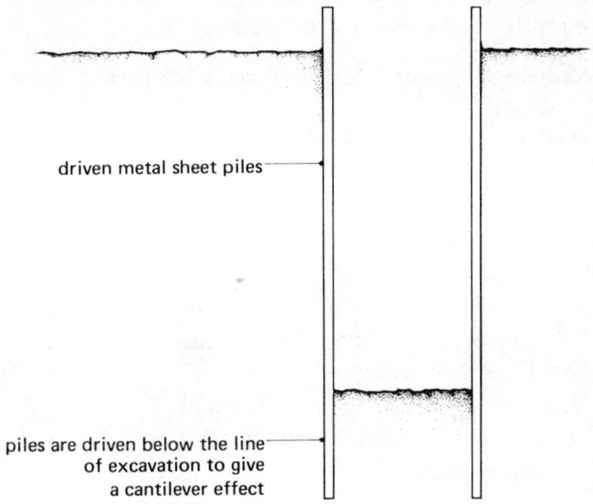

Figure 35 *Supporting trenches with driven sheet piling*

Scaffolding

This is covered on pages 102-17 as a separate part of temporary works.

Excavation

All work below ground level is dangerous. Excavation work usually indicates the commencement of work but, sadly, it can equally become the termination of life for an employee. Any one, or a combination, of the following can cause accidents:

Unknown soil structure
Lack of care in unstable soil
Inadequate protection
Poorly secured protection
Lack of care during weather changes
Insufficient inspections
Inexperience of supervisors and workforce

The principle of trench support is to hold back soil in a safe position thus enabling work to proceed without danger. The operative's head is the obvious, but not the *only* vulnerable point. If the soil collapses and encases the body, the rib cage can be crushed, squeezing the lungs and preventing breathing. The victim can die, even though his head is clear of the soil. The causes of soil collapse are numerous. A detailed study of these causes is essential. Soil collapse can be caused by:

Mechanical failure of soil unable to hold its own weight
Mechanical failure caused by change in soil consistency, brought on by rain or frost
Mechanical failure due to proximity of a previous soil movement or excavation
Soil movement caused by variations of structure, e.g. sand pockets or underground water ducts
Soil movement caused by vibration of moving vehicles and plant
Overloading at the edge of the trench
Impact of the soil, or its support by moving equipment or materials

Timber trench supports

The choice of timbering varies according to the nature of soil structure. A strong cohesive soil, e.g. clay, may only need intermediate supports, known as 'poling boards'. This method is called 'open poling' or 'open boarding' (Figure 36). The principle of all trench supports is for poling boards to be placed vertically against the excavation side. These are held in position along the trench by 'walings'. The strength of the support derives from the struts which are fixed tightly across the trench between the walings. These struts were traditionally made of timber, but in modern practice are likely to be adjustable steel props, which can be tightened to the exact requirements. Figure 35 shows various other blocks, wedges and props which are subsidiary supports. The timbering of trenches takes place as the soil is being removed, once the area to be supported is known.

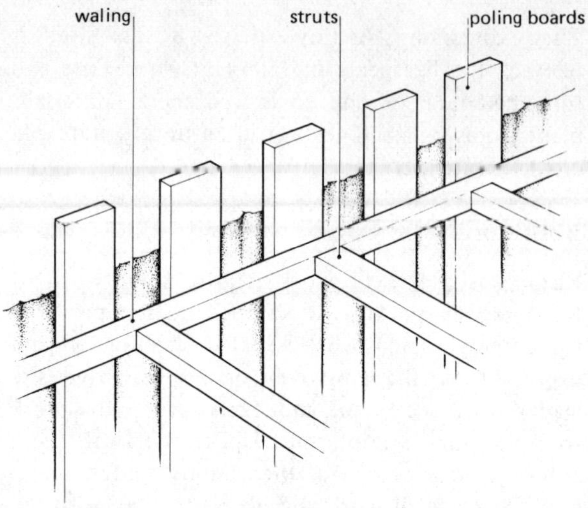

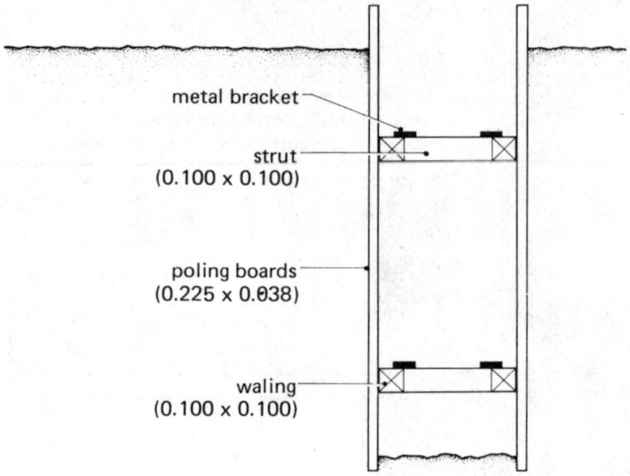

Figure 36 *Supporting trenches with timber*

Steel trench supports

The properties of timber have themselves led, in some cases, to unreliability, especially in loose, soft, or waterlogged soil. By inserting sheet steel supports before excavation, the ground around the trench can be held (Figure 35). Interlocking steel piles provide the best support for complete trench sides. A pile hammer (a weighted hammer driven by a compressed air ram) or a pile master (a suspended weight pushing down hydraulic jacks) drive the sheet piling into position. Special care is needed to ensure that underground service pipes and cables are not damaged or disconnected by piling. Good liaison with service engineers and local planning authorities are essential to avoid unnecessary expense, inconvenience and danger.

Battered sides

'Battered' sides are sloping rather than vertical, and therefore more secure. Where enough ground area is available around the excavation it may be practical to batter the sides of the excavation work. This is the safest procedure of work provided careful consideration is given to the batter required by the type of soil, i.e. its angle of repose.

Inspections of materials for trenches

All materials used for trench supports must be inspected every time they are used. Defective material must be discarded immediately. Testing must be completed by the inspection defined on pages 38-42.

Inspections of the workplace

Under the general requirements of the Construction (General Provision) Regulations 1961, all excavations exceeding a depth of 1.210 metres must be inspected every working day. These regulations also state that the last 2 metres of trench must be inspected before any work shift commences. A thorough examination recorded in Form 91 Part I Section B (see pages 38-42 for details) must be completed by a competent person every seven days, after the use of explosives and after soil or support movement. Inspections should check for:

Timber movement caused by drying-out or general timber defect, e.g. splitting
Ground movement, causing shrinkage and loosening of timbers
Soil movement from behind supports, causing slackness and subsequent collapse
Additional ground moisture swelling and displacing timbers
Displacement of timbers through impact with materials or plant

Protective fencing

There is a danger of site operatives falling into excavations. Barriers must be erected around falls of more than 2 metres. Although not a legal requirement, it is good practice to place barriers around every excavation on site. No specific ruling governs the strength or stability of these barriers. They just need to control the movement of site operatives. Tubular scaffolding is frequently used as a rigid barrier. Areas can also be roped off, but this system offers no resistance to movement. A clear line of demarcation is necessary to avoid any falls during the hours of darkness. During the winter months it is necessary to illuminate the whole area. Warning lights are necessary after working hours if the work is close to a public thoroughfare or if there are night watchmen or security patrols on the site.

Ground water

The soil structure contains natural ground water up to a certain level, the 'water table'. Its level within the soil fluctuates during dry and wet seasons. When excavations penetrate into this zone, it becomes necessary to lower the water level, which can affect the supporting members. This procedure is called 'de-watering' or 'ground-water lowering'. It is generally a specialized exercise, particularly on larger contracts. The simple form of de-watering is to make well points or sumps from which the excess water is pumped clear. Drained soil is considered to be more stable than wet soil, but the safety procedures are still necessary.

General considerations

Work below the ground creates hazards which are covered in detail elsewhere in this book. Here is a simple check-list of hazards:

Machinery falling into trenches – pages 125-33
Safe means of access and egress – pages 118-24
Head injuries from lowered materials – page 96

Checklist for trenches

A site leader should consider the following fundamental checks on excavations:

Before work begins

Find, locate and mark public services
Liaise with Safety Officer and appoint competent person
Organize plant, and required working space

Organize delivery and inspection of support materials
Provide protective clothing, garments and appliances
Provide suitable barriers
Organize and arrange inspection of ladders
Notify all public services (police, traffic control, and bus companies), etc. of all work that may cause restriction or danger

While work is in progress
Liaise with safety officer and competent person regarding inspections and suitable record-keeping
Organize a balanced workforce – avoid overcrowding in a trench
Arrange adequate fencing, lights and warnings around the excavation
Arrange safety stops for all site transport near trench areas or excavations
Check protection of adjacent buildings and safety hazards to traffic
Install a safe evacuation procedure
Check regularly for 'unseen hazards', e.g. noxious fumes or gases
Plan and prepare for safe backfilling activities
Maintain tidy work area at all times

Shoring

This term is given to temporary works which support buildings during reconstruction or presenting danger. There are three basic types of shoring schemes:

Dead shore

Horizontal members called needles pass through the building internally and externally. These needles bear down directly on to vertical struts which in turn fit on to sole plates at ground level. Once this support has been achieved (Figure 37) the area beneath can be renewed, repaired or renovated.

Raking shore

Timbers propping an insecure wall at an oblique angle. The supporting 'rakers' bear down on to a timber sole plate or grillage of timbers. The top joint forms an integral bond between the rakers, a wall support timber and a needle passing through

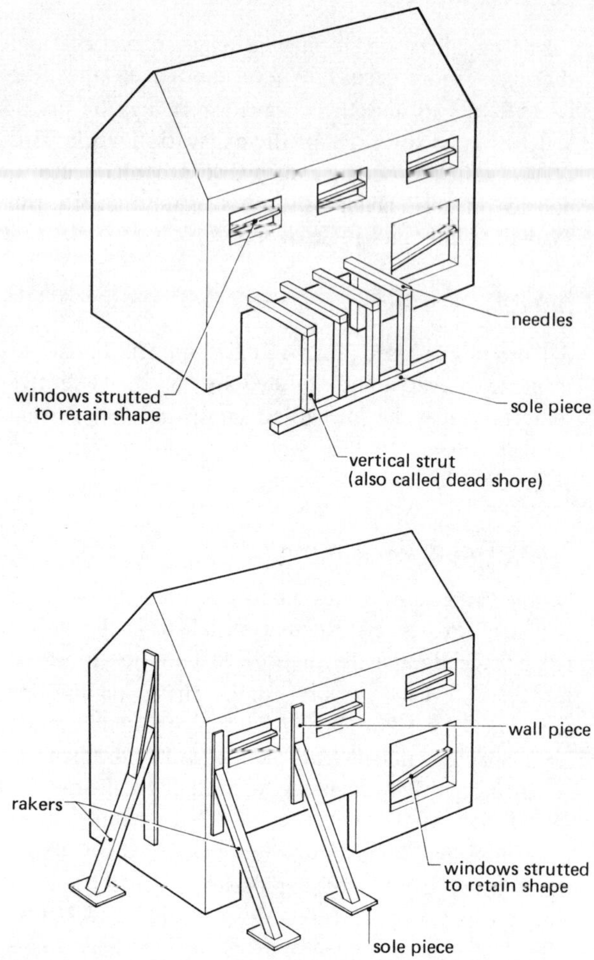

Figure 37 *Supporting a building with dead and raking shores (for further details see R. Bayliss, Carpentry and Joinery 3, 2nd edition, Hutchinson, 1969)*

the wall. The stability of the temporary support comes from the intersecting centre lines of each member. Additional vertical propping may be necessary between floors in conjunction with the raking shore.

Flying shore

Shoring against other structures. Where it is impracticable to form raking shores, and where buildings opposite are structurally sound, a system of flying shores may be introduced (Figure 38).

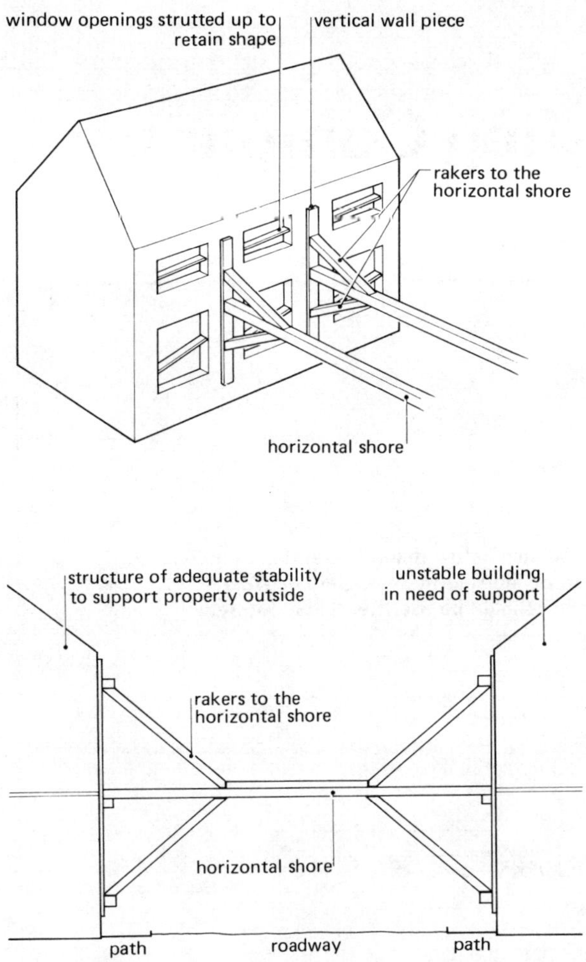

Figure 38 *Supporting a building with flying shores (for further details see R. Bayliss, Carpentry and Joinery 3, 2nd edition, Hutchinson, 1969)*

Inspections of materials for shoring

Irrespective of the type of shoring scheme, the materials – generally timber – need to be inspected before use. All defective materials need to be discarded. If steel adjustable props or similar steelwork are used a similar inspection must be made before use.

Inspection of the workplace

A complete inspection of the support system must be carried out every seven days. The following is a typical check-list of points to look for during inspection:

Timber distortion through excessive loading
Timber movement caused by drying-out
Movement of timbers and/or support system through soil movement, due to an increase or decrease of moisture in the ground
Displacement of timbers by moving plant or materials
Movement or displacement of timber connectors, bolts or straps
Movement of support system due to additional load from the building, e.g. possible settlement or collapse

Protection

A clear line of demarcation around the temporary shoring must be erected to avoid damage from plant or machinery. Similarly, site lighting must be installed if the shoring is near a public thoroughfare or if night watchmen or security patrols are operating on the site. Additional protection may be necessary to flying shores to indicate the clear headroom for site or public traffic.

General considerations

Shoring or similar temporary works create hazards relevant to other sections of this book:

Safe means of access and egress – page 118-24
Head injuries from falling objects – page 96
Inspections and records – pages 38-58
Body protection – pages 96-101

Before work commences

Liaise with safety officer and appoint competent person
Organize plant and space requirements
Organize delivery and inspection of support materials
Provide protective barriers and lighting
Provide protective clothing and equipment
Notify all public services (police, traffic control, bus companies, etc.) of all work that may cause restriction or danger
Prepare site surveys, including progress photographs as protection against damage liabilities

R & D Building Company

Pararad Road, Roselip

SITE:
CONTRACT NO:
DATE:
REPORT NO:

SITE INSPECTION REPORT

The numbered items marked in the column below require attention as indicated. These are considered to be a breach of the Health and Safety at Work Act 1974, and the Factories Act, and should be rectified immediately.

1 SCAFFOLDING
2 LADDERS
3 HOISTS AND LIFTING APPLIANCES
4 CRANES AND EXCAVATORS
5 SITE VEHICLES
6 PLANT, MACHINERY
7 PORTABLE TOOLS
8 TEMPORARY ELECTRICS
9 EXCAVATIONS, GROUNDWORKS
10 SITE TIDINESS
11 PROTECTIVE EQUIPMENT
12 L.P.G. H.F.L.
13 WELFARE FACILITIES
14 FIRST AID
15 FIRE PRECAUTIONS
16 REGISTERS
17 SECURITY
18
19
20
21 SAFETY AWARD Points / Comments

Copies: Site
Contracts Manager
Contracts Director
Safety Office

Signed........................ Safety Officer

Figure 39 *Form for site inspection report*

While work is in progress

Liaise with safety officer and competent person about inspections and record-keeping
Maintain adequate barriers and lighting
Organize safe storage and stacking of materials
Install a safe evacuation procedure
Plan and prepare for safe dismantling
Maintain a tidy work procedure at all times

A typical site inspection report is shown in Figure 39.

10 Body protection

As previously summarized (pages 63-71), there exists a demarcation line between 'health hazards', which cause damage to internal body organs, and 'physical hazards'. Where there are physical hazards, body protection is required to avoid or minimize the risk. There are responsibilities from both employer and employee which have been shown clearly in chapters 2 and 3. This chapter therefore reviews hazardous situations that create risk to the employees' body and show how to reduce or eliminate the risk.

Protection for the head

Employees engaged upon normal site work can be protected by wearing a safety helmet. These are manufactured in two parts. The protective shell is moulded from thermosetting plastics which form a hard durable cover over the head. Within this shell is fixed a flexible skull cap. A clear gap must be maintained between the shell and the skull, in accordance with BS 5240. The inner skull cap is adjustable to fit any size of head and to allow the user complete comfort. There are several styles of protective helmet, each with merits and demerits, but all reduce impact from falling objects. The legal controls do not enforce the employer to provide safety helmets. Although helmets are not specifically listed, the legal powers require that the employer must provide protection so far as reasonably practicable. To achieve this and offer good safety practices, most companies engaged in construction provide helmets for use at work. According to the company safety policy, there may be an enforcement regarding the issue and use of helmets. The following forms good safety practice either individually or in combination, according to company policy:

Rule within the safety policy that helmets must be provided and worn on all sites or where danger to employees' heads might exist.

Ensure that enough helmets, in good condition, are provided for everyone employed on the site and that they are easily available.

As an improvement to the preceding point, issue each employee with his own helmet. This avoids employees having to wear helmets that may have been previously misused.

Issue adequate advice and warning to employees regarding helmet use.

Enforce the use of helmets, with supervisors checking their use; an incentive scheme, where employees receive a bonus for constant use of helmets, will help.

The employee is obliged to take reasonable care. This simple phrase constitutes a moral and legal requirement to wear protective headgear whilst at work.

Dangers to the hands

Suitable protection to hands will vary according to the trade or activity of the employee. Hazards relating to skin infections are considered later in this chapter. Other dangers to the hand are as follows.

Burns

Burning of the body skin caused by the contact of a flame or enflamed material in natural working conditions, not as a result of fire, are considered here. Welding creates a spontaneous ejection of burning or partially burnt metal. Where this falls on to combustible materials it continues to burn.

If the burning metal falls on to the hands of the welder he will suffer superficial burns and this must be avoided (see also pages 207-9).

Chemicals

Numerous chemicals and gases burn. The hands are again vulnerable to abuse from careless use of these liquids, e.g. paint-strippers and acids.

Poisons

When a poisonous substance is absorbed through the pores of the skin it can enter and contaminate the bloodstream. This problem exists where lead is consistently used. The same situation applies, albeit with reduced hazard, where lead-based materials, e.g. lead paint, or impregnated timbers, are used.

Contamination

The hands have numerous vessels and chambers, e.g. under the fingernails, where poisonous substances can be retained, especially if hygiene is not maintained. Particular problems of contamination exist where asbestos rope or other fibrous material is used.

Impact

Impact upon the hands can cause discomfort and/or disfigurement. No protection exists to avoid the impact, but carefully chosen gloves may reduce the effect and the pain inflicted.

Protection for the hands

Employers are bound by statutory control and have a moral obligation to provide protection so far as reasonably practicable. This applies particularly to those working on known hazardous materials.

Burns

Leather gloves should be used at all times to protect the employee from molten or burning metal, sparks and general heat distribution from welding, etc. The gloves used need to be sealed at strategically placed seams and to be sufficiently long to protect the complete hand and forearm.

Chemicals

Without digressing into the intricate details of chemistry the full realm of protection is difficult to cover. The industrial chemicals most used within the construction industry are paint-strippers and liquid solvents. Certain acids are used in liquid form but, more frequently, as catalysts within an adhesive compound. Where hazards exist, suitable rubber gloves must be worn. The thickness of the rubber, its durability, and the area of protection will depend upon the severity of the acids used.

Poisons

The types of work where poisons can be absorbed directly into the skin pores are numerous, including chemicals or lead-based materials. Gloves cannot be worn if these materials are to be used competently and if output is maintained, so hand barrier creams should be applied prior to working with toxic substances. This protective cream retards or eliminates absorption through the skin. It is also important to wash one's hands thoroughly prior to food breaks and before leaving work.

Contamination

Where fibrous materials are handled, rubber gloves must be worn to avoid contamination of the hand. With milder contaminators hand barrier creams will usually be sufficient protection.

Impact

Rough handling activities, or exposure to sharp edges of materials and equipment need a general purpose glove usually made with a rubber palm and cloth backing. BS 1651 details the specific requirements of industrial gloves. The degree of protection must be proportionate to the hazard to be eliminated.

Dangers to the eyes

Eye malfunction can be caused through exposure to intense visible light, e.g. ultra-violet rays and infra-red rays given off by welding processes, or by solid matter hitting the eye and damaging its active parts. Liquid of hazardous or burning character and dust may also affect the vision of the eye.

Protection of eyes

The Protection of Eyes Regulations 1974 immediately followed the Health and Safety at Work etc. Act and made provision for the protection of eyes of the employees in several specific areas of construction work. The protection required for certain types of hazard are as follows:

Impact of solids

The anticipated velocity of the impact is an important criterion when deciding the eye protection to use. BS 2092 established the standard of lens to be used in goggles and classified these as:

Grade 1 – to withstand the impact of a standard 6.5 mm steel ball at a velocity of 119 metres/second

Grade 2 -- to withstand the impact of a standard 6.5 mm steel ball at a velocity of 45.7 metres/second

The lens is however dependent upon the supplementary support of a good frame and adequate means of ventilation to avoid misting up. Provided goggles with the appropriate lens are worn properly the employee will recieve reasonable protection. Nevertheless there must be serious reservations regarding the graded styles available. A particular hazard to be avoided is where different grades of goggles are acceptable in adjacent work areas. An example is if a painter's workshop, requiring Grade 2 goggles for stripping, is sited next to a plumber's workshop, requiring Grade 1 goggles for high-speed grinders. An innocent change or borrowing of goggles may protect a painter and endanger a plumber. To eliminate this, all operatives must be advised of the lens marking (at the front, top edge), or a colour coding of workshop equipment can be implemented.

Impact of liquids and dusts

The main hazard from liquid is not the velocity of the impact nor the intensity of materials hitting goggle lens. The requirement is that operators' eyes are protected from hazardous chemicals and dusts, by having a goggle fitting closely to the contour of the face. There are no strict lens strength to examine and BS 2092.B goggles are usually acceptable.

Blindness – glare

The high degree of intense light and infra-red rays given off by welding and similar hot metal processes can cause serious eye damage. The hazard is not immediately appreciated but continuous exposure over a period of time will cause damage. Goggles with a lens which prevent strong light from transmitting to the eye will give protection. Details of suitable goggles are given in BS 679. A good degree of protection can be achieved by the use of head shields which combine a facial shield and eye protection in one easy-to-use apparatus. (See also page 208.)

Where a person is employed upon a process that may endanger his sight, the employer must provide adequate eye protection. In some cases shields for use in conjunction with goggles are required. The employee should be issued with his own equipment for full use during his work activities. Where employees may be at risk by working adjacent to the hazard, but are not necessarily participating in the activities, a personal issue must be made to eliminate their risk. Whenever a process is to be carried out that will cause risk to the employee, there must be suitable eye protection available for the duration of that process. Under no circumstances is any employee to work periodically or permanently at a hazardous job without eye protection. All lost, damaged or inadequate goggles must be replayed by the employer at the employee's request.

Markings

The following table outlines the markings applicable to eye protectors as required for several construction industry processes.

Type		*Lens marking*	*Housing marking*
general purpose		(1) BS 2092 (2) manufacturer's mark or licence number	(1) BS 2092 (2) manufacturer's mark or licence number
chemical		*as for general purpose*	*as for general purpose* and letter 'C'
dust		*as for general purpose*	*as for general purpose* and letter 'D'
gas		*as for general purpose*	*as for general purpose* and letter 'G'
impact	Grade 1 Grade 2	*as for general purpose* and the figure 1 or 2 *	*as for general purpose* and the figure 1 or 2
molten metal		*as for general purpose,* additionally marked with letter 'M'*	*as for general purpose,* additionally marked with letter 'M'
combination impact and/or molten metal with chemical, dust or gas		*as for impact and/or molten metal* *	*as for impact and/or molten metal,* and letter 'C', 'D' or 'G' as appropriate

* Glass lenses supplied for use in impact and/or molten metal eye-protectors in combination with plastic lenses are additionally marked with the word 'outer'.
Reproduced by kind permission of the British Standards Institution.

Dangers to the feet

There is no legal control to enforce the employer to provide suitable footwear. Where, however, he expects work to proceed in wet conditions or special footwear is essential to do the normal work, then he must provide it. He should however advise and recommend that all employees have good foot protection and many firms arrange cost-price purchase for their employees. In some companies, the employer provides safety footwear, a good practice which tends to be an exception rather than a rule. There are two areas of hazard to the feet:

Impact

A load falls on to the top of the foot, causing crushing or fracture.

Puncture

A sharp protruding object can press through the footwear and sever or break the skin of the foot.

Protection for the feet

The employee is responsible for his own footwear other than the provisions just mentioned. The two hazards can be eliminated either individually with separate styles of boot or concurrently by the purchase of dual-purpose safety boots.

Steel toe caps

To eliminate the impact on top of the toes a boot is available with steel toe caps incorporated within the manufacture. There is no excessive weight increase and appearance is not impaired.

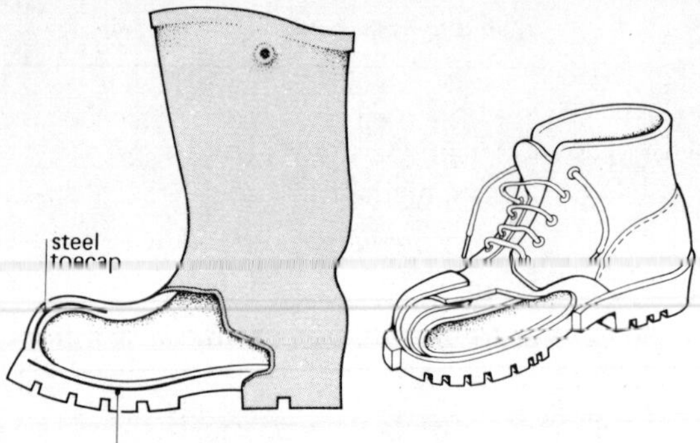

Figure 40 *Safety boots. The boot incorporates a steel toe cap, and sometimes also flexible steel soles*

Flexible steel soles

To resist the protrusion of sharp objects through the sole of the boot a flexible steel sole is built into the sole of the boots. The heel, being adequately thick, needs no additional strength. No weight increase is apparent and the flexibility of the steel does not restrict the manoeuvres of those wearing them.

Combined steel protection

Certain manufacturers have now manufactured an industrial safety boot incorporating both steel toe caps and flexible steel soles. These combine complete safety with little increase in weight, no restrictions and an acceptable appearance.

Skin diseases, infections and damage

The intention of this section is to create an awareness of problems arising from skin infections. Cuts and grazes have been included previously (page 96). Burns will be reviewed in specific chapters where the hazard exists.

Skin problems

The problem varies from person to person, according to their hygiene and the sensitivity of their skin to the materials or substances used within construction. Certain types of employment cause differing levels of exposure to skin dangers.

Dermatitis and infections

Many fluids, chemicals and materials are harmful to the skin. Medical treatment is usually necessary to clear the problem, and much discomfort can be experienced during the cleansing period. In some cases scars remain permanently or for prolonged periods after the problem has been cleared.

Damage

Construction workers are frequently exposed to tasks that have little regard for the employees' skin. A supplementary problem is the stupidity of colleagues who persist in joking about those who wear hand protection, or face shields. The most vulnerable areas of the body are those not normally covered by clothing – hands, forearm and face. These areas of the body are important to one's appearance and deserve protection.

Skin irritations

Although these do not cause serious markings or scars they are uncomfortable and unpleasant. The irritation, if allowed to persist, is the preliminary stage of inflammation which may develop into a skin disease.

Protection from skin problems

The degree of protection must allow employees to be safe and able to work. Petty restrictions should not become a prelude to confrontations with management. There have always been minor operations that require action rather than a rule book, law or vast protection scheme. Remembering these qualifications, here are some guides for personal protection.

Personal hygiene

The oldest, most simple form of personal cleanliness is to have a good wash immediately before taking a food break and at the end of the working day or shift. Soap and water will suffice as a barrier against minor infection. Of course, serious hazards exist that need more substantial protection but harm from operations that are simply dirty can be counteracted by good washing. The employer must provide adequate washroom facilities. These consist, generally, of one basin or washing vessel per twenty-five persons where the work is for longer than six weeks (see page 175 for details). The employee's responsibility is to use the facilities provided, and not to ignore them or abuse them by dirty habits that may discourage others from using washing facilities.

Protective creams

Several creams are now available for industrial use under the broad heading of 'barrier creams'. In dirty working conditions, the employer should make available this form of protection. The face or hands are covered with a thin layer of this oily substance containing an antiseptic treatment, to deter the absorption of dirt or germs. Ordinary washing easily cleanses the skin of the cream and any superficial dirt.

Cleansing creams

Where face or hands have been subjected to dirt or any similar hazards, creams are available with antiseptic cleaners. Employees should realize that this is only the first part of a cleaning operation and the cream must be washed off with water afterwards to prevent a drying of the pores which will allow infection into the skin structure.

11 Scaffolds

A scaffold is a temporary platform provided for work activities and the storage of materials. It may be above or below normal ground level. The scaffold also incorporates any access to, or egress from, the working platform.

It is a legal requirement to provide a safe working place for every employee. The scaffold working platform is one of the most common sights at any construction area and, at the same time, is one of the most legally abused working provisions. Current statistics confirm that most fatal accidents occur as a result of a fall from scaffolding. These two points, considered together, imply that scaffolds are the cause of most fatal construction site accidents. This chapter outlines the fundamentals of scaffolding and therefore does not extend into all the various styles and imperfections of scaffolds manufactured. Several companies produce temporary working platforms, each having merits and de-merits. It is not the intention to promote or denigrate any form or type but to provide a general review of scaffolding procedures.

Definition of terms

The following brief terms applicable to scaffolding will aid the understanding of this chapter.

Base plate – a flat metal plate of 0.155 m sides incorporating a central pin to which standards are located on to the sole plate. Holes are also provided to secure the base plate to the sole plate.

Brace – a diagonal tube fixed to other members to create stability by triangulation.

Bridle – a dummy putlog without the flattened end for use where the structure cannot give support, e.g. door and window openings.

Couplings – metal threaded connectors, in accordance with BS 1139, for joining tubular members, e.g. putlog-coupler; swival-coupler; double-coupler.

Guardrail – horizontal tube member to the top perimeter of the scaffold to prevent falling from the platform.

Joint pin – a short, internal fitting for joining tubular members end to end.

Ledger – a horizontal tube joining together standards and supporting putlogs or transomes.

Puncheon – a short standard that is not seeking support from the structure or ground, e.g. an additional scaffold to a hoist platform.

Putlog – a horizontal tube with one end resting upon a ledger and the second end flattened and located into the structure of the supporting building to support scaffold boards.

Reveal tube – a short tube, secured between window reveals for stabilizing the scaffold structure.

Scaffold lift – a height between ledgers that is worked up the building, usually 1.520 m.

Sole plate – a substantial section of timber, resting upon the ground to provide an even distribution of load and avoid ground irregularities.

Standard – a vertical tube member supported on a base plate which rests upon a sole plate.

Toe board – horizontal board, on edge, along the platform to retain loose materials or tools.

Transom – horizontal tubes, supported by ledgers, and supporting scaffold boards.

Tube – a random length of mild steel, 0.048 m outside diameter, conforming to BS 1139. The requirements are that defects, surface flaws, or similar visual deficiencies are not permissible, and all end cuts must be true and square to the tubular axis.

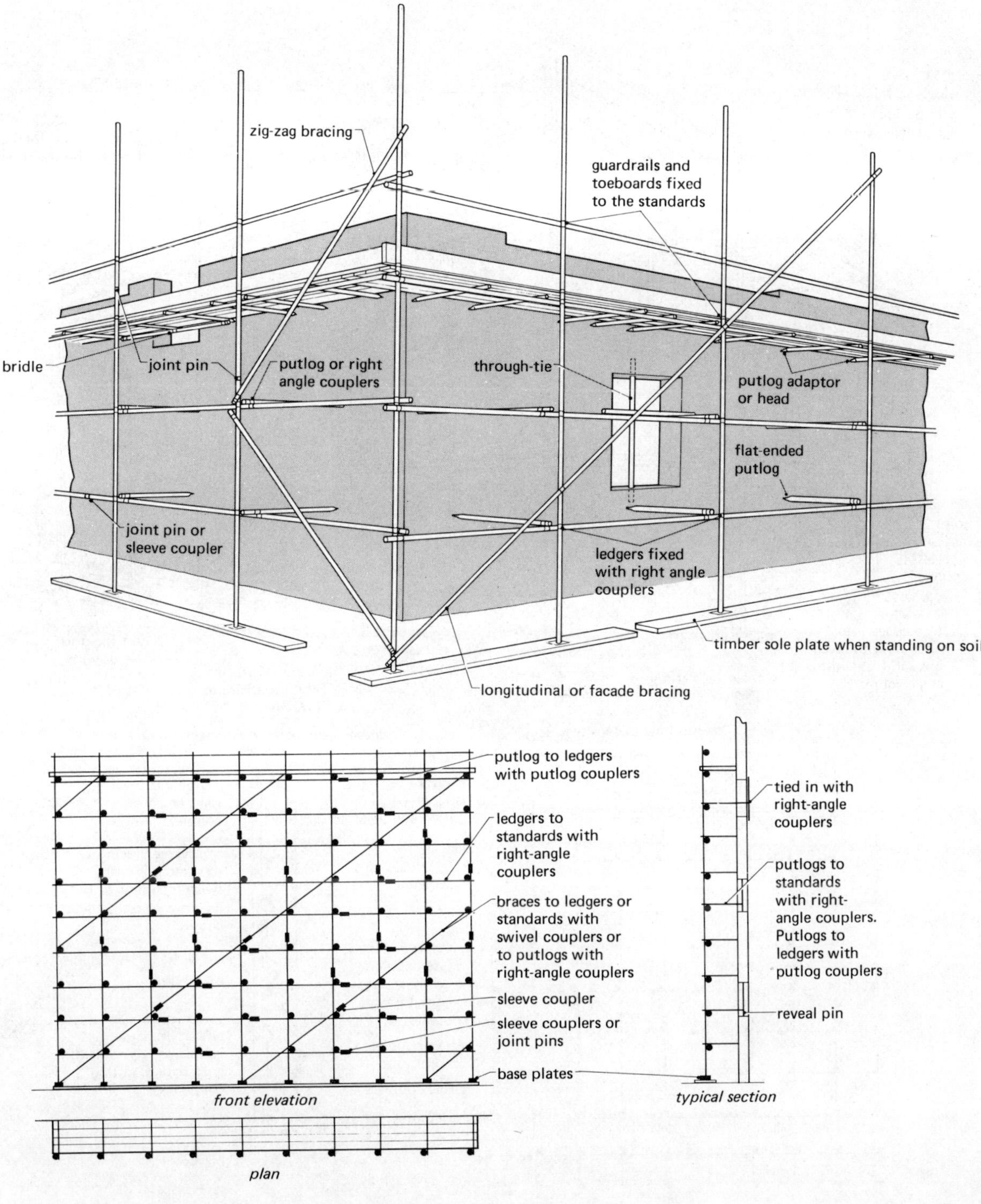

Figure 41 *Putlog scaffold – view, plan, front elevation and section*

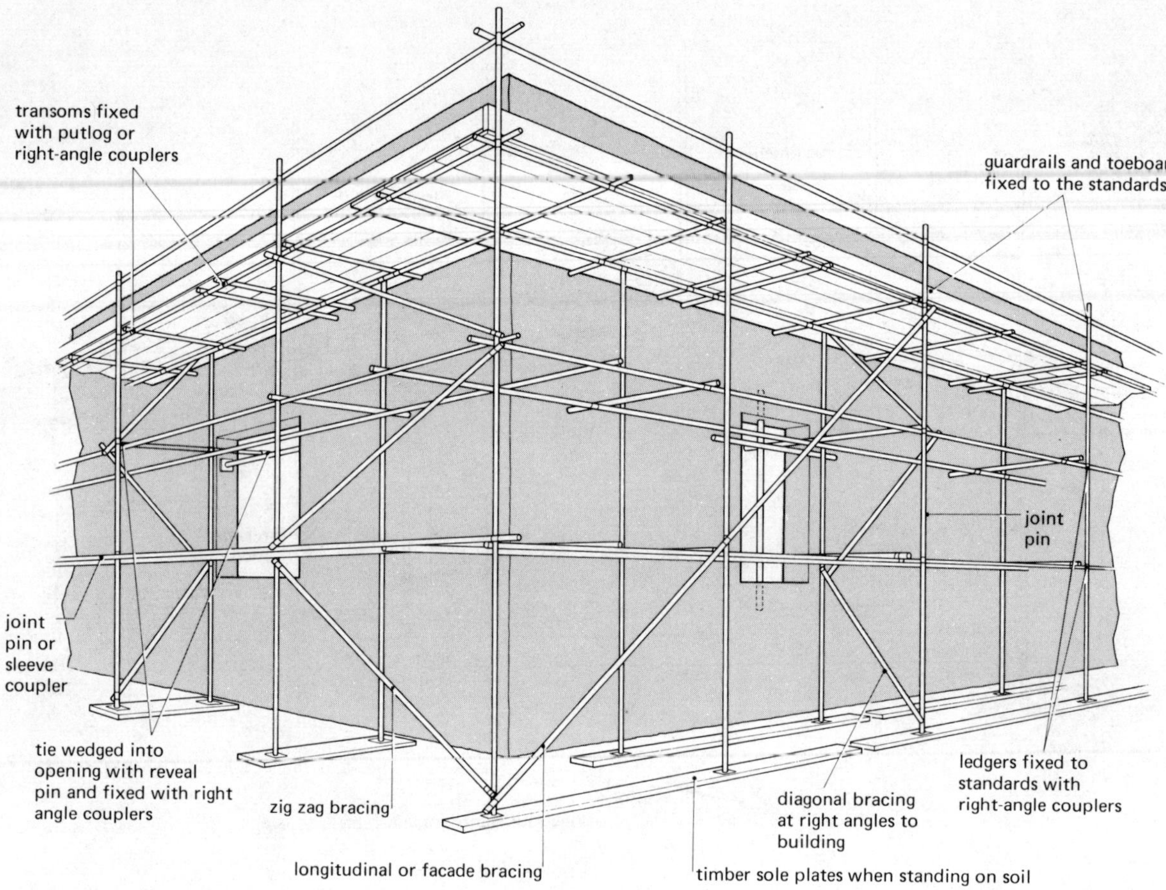

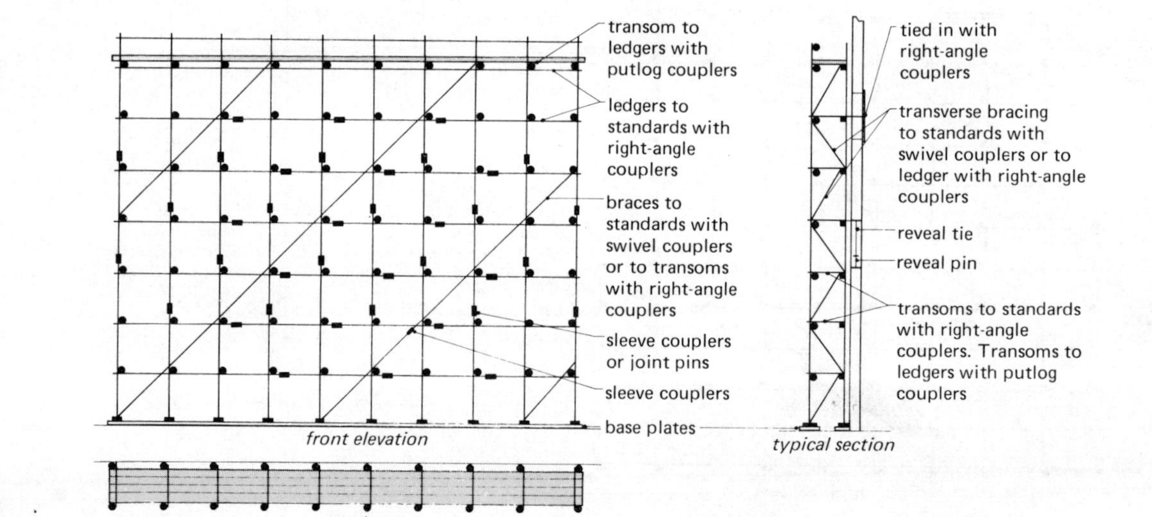

Figure 42 *Independent scaffold – view, plan, front elevation and section*

Scaffold types

Putlog scaffold (Figure 41)

This is a scaffold in which the building gives direct support to one complete elevation of the scaffold unit. A single row of standards supports horizontal ledgers, which in turn support, with the assistance of the building structure, the horizontal putlogs. These putlogs bear the loading of scaffold boards to create the platform. The use of this scaffold is restricted to places that allow the insertion of putlogs into the fabric of the building, e.g. brickwork and certain masonry buildings. Non-load-bearing or decorative building facades are not suitable for putlog scaffolds.

Independent scaffold (Figure 42)

The name implies an independence, but this applies only to the load-bearing capacity of the scaffold structure. The building is still used to control any movement of the scaffold and prevent it moving away from the building facade. Two rows of vertical standards support two horizontal rows of ledgers, which in turn hold transoms across the width of the scaffold. These transoms are the load-bearing supports for the boarding which creates the working platform. This scaffold is frequently used for work to existing buildings when the non-load-bearing characteristics of the building render it impossible to gain support, e.g. curtain wall construction or timber frame developments. It is sometimes known as 'Masons' scaffold.

Interlocking modular scaffold

This is a temporary structure of modern type used where 'tubular' members would be too cumbersome and restrictive. It is quickly erected and offers a good interchange of standard units which are assembled without clips and with a minimum of loose components. It is a completely independent system for load-bearing purposes and relies upon some support from the building for vertical stability. The uses of this scaffold are similar to the independent type. It can be used for complete support around an entire building or as a small platform to a short elevation. In principle, the interlocking scaffold consists of a series of standard units which interlock without clips or any other fixing agents.

Tower scaffold (Figure 43)

This is a tall unit with a small base area. It can be prepared with tubular steel members clipped together or with interlocking members. Its main uses

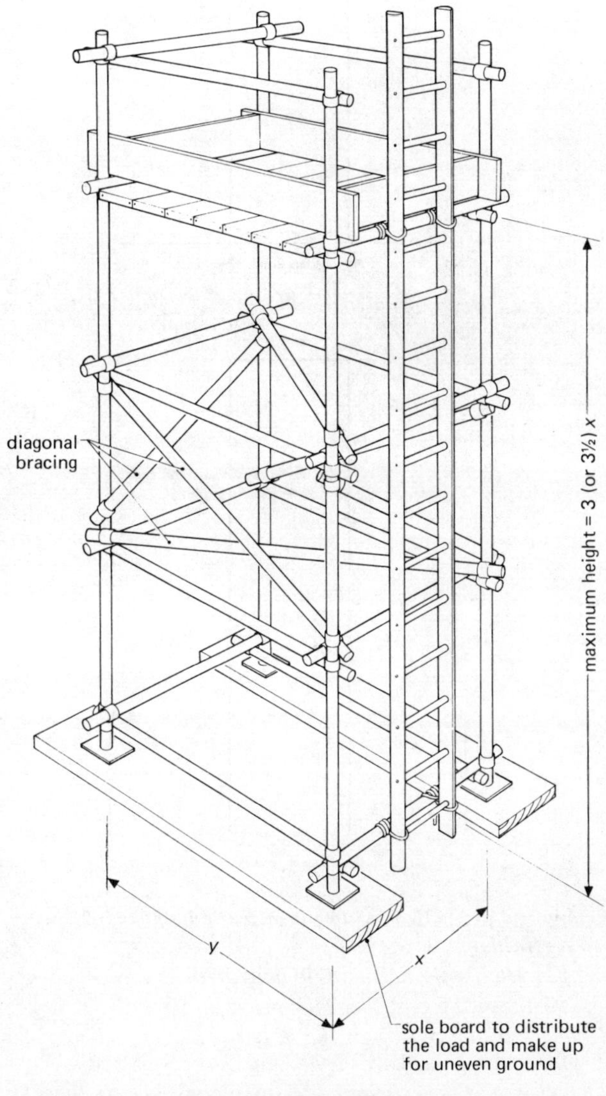

Figure 43 *Details of static tower scaffold. Maximum height is 3 (or 3½) times x, the length of the short side, not y, the length of the long side*

are for individual work areas where isolated work is needed, e.g. to reach individual windows or chimney stacks. There are certain limitations upon this arrangement with regard to the base size in relation to total height (see page 105).

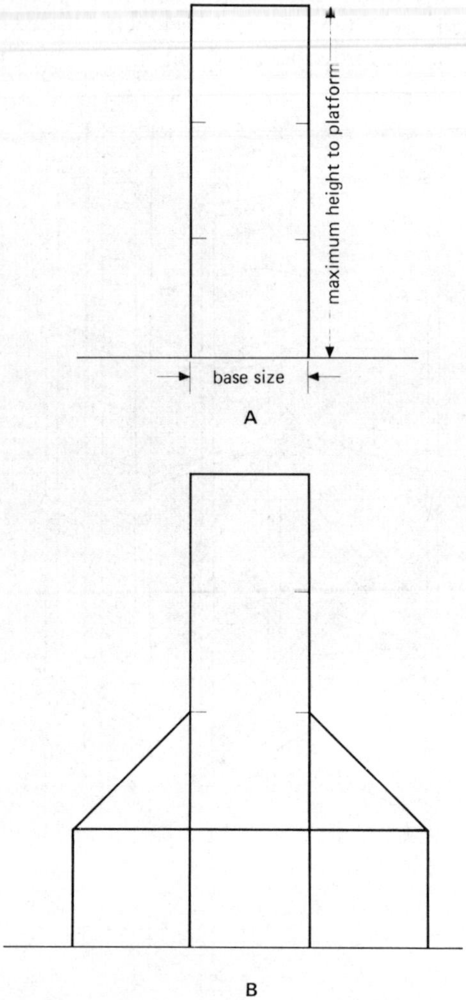

Figure 44 *Maximum permissible heights of tower scaffolds.*
A – Maximum permissible height for external work is 3 times the base size (for internal work the maximum is 3½ times the base size)
B – If a higher tower is needed, outriggers give increased base size and additional stability. If the base of the tower is not square, the maximum height is 3 (or 3½) times the length of the shorter side (Figure 43)

Mobile tower scaffold (Figure 45)

This is similar to the preceding type but has the added advantage of mobility. It allows access to individual work locations, and is used on work repeated at a constant height, e.g. for several windows, light fittings or isolated ceiling repairs.

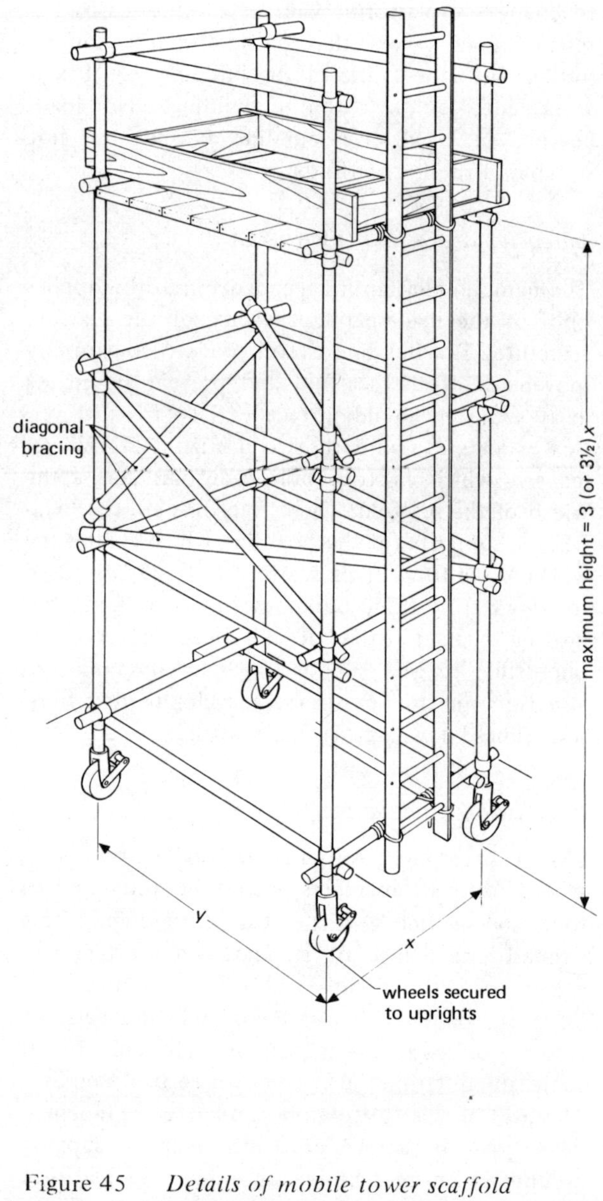

Figure 45 *Details of mobile tower scaffold*

Trestle scaffold (Figures 46 and 47)

This is intended for light work of short duration, e.g. decorating or lightweight panelling. In Scotland it is frequently used for internal brickwork. The constituent parts are two pairs of trestles to support the boards or staging to form a working platform. Certain limitations are enforced to restrict the use of these (see page 115).

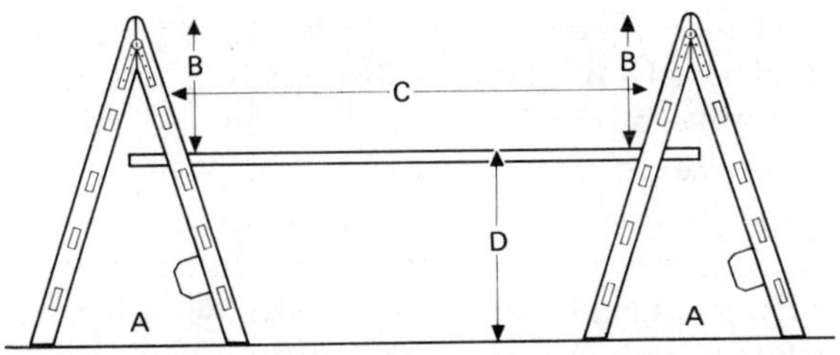

Figure 46 *Permissible tolerances for a trestle scaffold.*
A – Trestles must always be stood on a firm, level base
B – Safe working procedure keeps at least the top one-third of the trestle clear above the working platform to retain stability and act as a holding rail
C – Maximum permissible spans of boards are : 1.300 m for 0.040 m boards, 2.440 m for 0.051 m boards. Proprietary staging is preferred; its maximum span is 3.048 m. At least two boards wide must be used
D – A trestle scaffold must not be used where anyone can fall more than 4.570 m

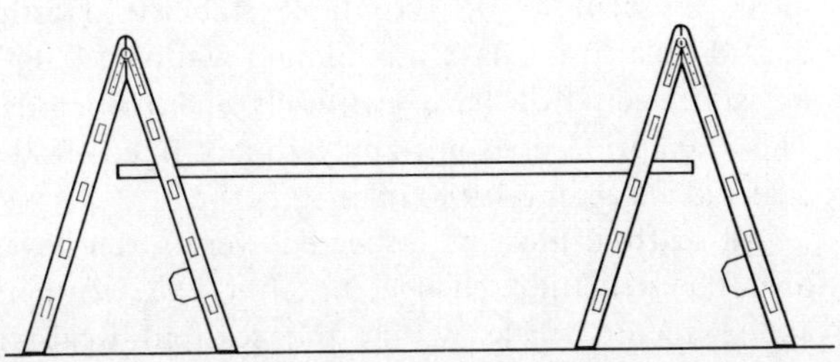

Figure 47 *Trestle scaffolds can be used at ground level or upon other scaffold platforms*

Special roofing scaffolds (Figure 48)

These are required to prevent falls from the roof surface. There are two types.

Within new construction, or during substantial repair works, an adaption to the normal scaffold is recommended (page 115).

On smaller work activities, it is acceptable to use a secured ladder and crawling board (page 108).

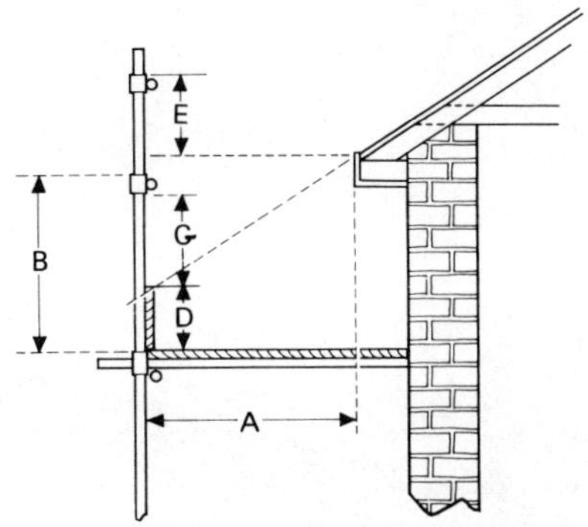

Figure 48 *Adaptation of scaffold for roof work.*
A – Minimum scaffold width is 0.640 m
B – The guardrail is between 0.920 and 1.150 m high
C – The maximum gap between the toeboard and guardrail is 0.760 m
D – The toeboard rises to the line of the roof slope and is not less than 0.155 m high
E – This distance does not exceed 0.440 m, and the distance between guardrails does not exceed 0.760 m

Legal interpretation

The main area of legislation will be dealt with later (pages 166-76). The following particulars are applicable to scaffolding.

The Construction (Working Places) Regulation 1966

This regulation deals with the working places of employees and has the control of scaffolds as its main function. The document is divided into four parts in which thirty-nine sections deal with working places:

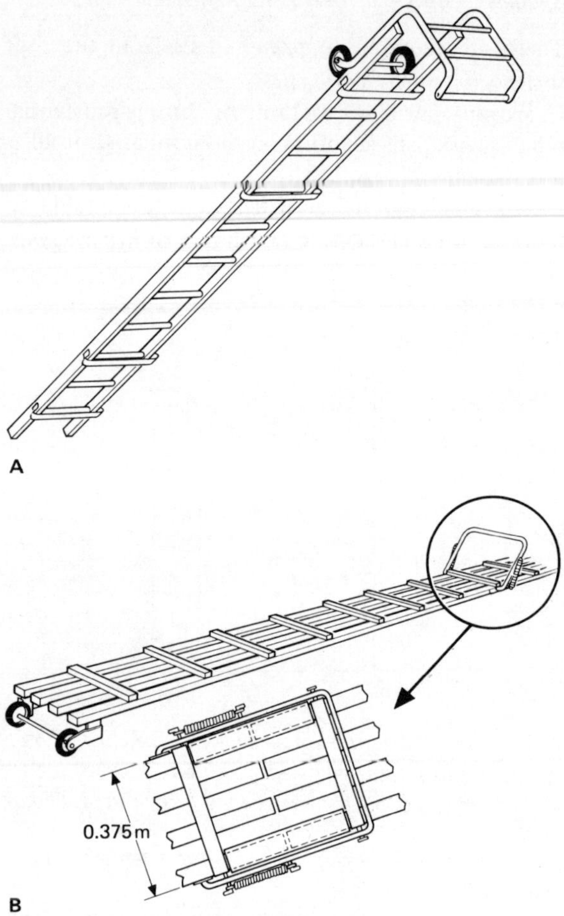

Figure 49 *Roof ladders.*
A – Metal roof ladder fitted with wheels to assist in placing in position
B – Wooden roof ladder designed to be fitted over the ridge and to have more sections added

Part 1: Application and interpretation – all details of application and interpretation are included with definitions of employers and employees and an outline of their obligations to each other. Brief interpretations are included to qualify the terminology, e.g. plant or equipment includes scaffold boarding as non-mechanical plant, and a working platform is a place of a man's work.

Part 2: Exemptions – certain exemptions apply, by consent of HM Chief Inspector of Factories, e.g. for steeple jacks, provided safety is maintained and compliance is not practicable.

Part 3: Safety of working places and access and egress – details of ladders, their fixture, and general routes to the working place are detailed, e.g. the angle of ladder is to be 75° and the minimum rise of ladder above working platforms is 1.070 m. Scaffolds must be provided where work is inaccessible and be strong enough with acceptable quality materials in accordance with BS 1139 Metal Scaffolding and BS 2482 Timber Scaffold Boards. No defective materials are to be used, and suitable maintenance shall be arranged to avoid damage to or displacement of any scaffold parts.

Any partially dismantled scaffold must be left safe, and bold notices erected, advising of the state of the scaffold. Access to unstable areas must be blocked.

All scaffolds must be erected to a minimum standard: sole plates and base plates must be used; ledgers must be well spaced for stability, putlogs must be well located; and spans of boards must be observed (for 0.040 m planks the maximum span is 1.520 m; for 0.051 m planks the maximum span is 2.590 m).

Any mobile scaffolds must be locked whilst in use, and when exceeding 9.750 m high they must be secured to the building.

The maximum height of a scaffold blocked from the ground direct is 0.610 m (Figure 50).

The maximum permissible height of any trestle scaffold is where anyone can fall more than 4.570 m. Where trestle scaffolds exceed 3.660 m high they must be secured for acceptable stability. Trestle scaffold platforms have a minimum width of 0.440 m, and need not have guardrails and toeboards. This exemption does not apply to any fixed trestle scaffold exceeding 1.980 m high.

All scaffold must be inspected every seven days, or after structural change, or after bad weather, and a record made within the scaffold Register Form 91. All inspections must be completed by a competent person. Any scaffold less than 1.980 m high is exempt.

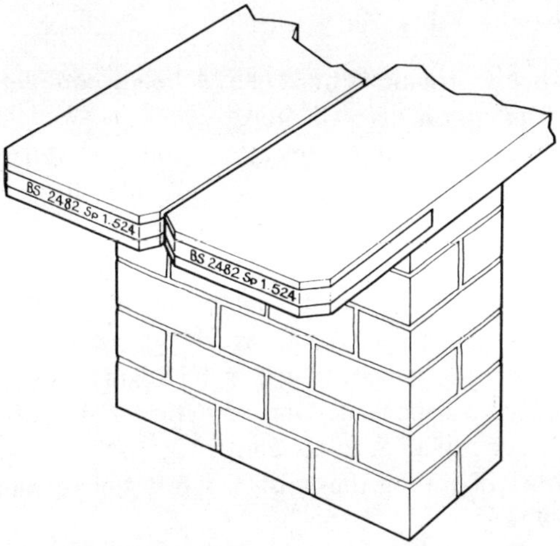

Figure 50 *The maximum height for blocking direct from ground level – using standard 0.038 m scaffold boards blocked up on building blocks or similar material, suitably bonded – is 0.610 m*

All scaffolds must be close boarded with a maximum slope of one to one and a half. Boards must be of 0.210 m minimum width if 0.051 m thick, and 0.155 m wide if more than 0.051 m thick. Board ends should not overlap any support by more than four times the board thickness. Where boards overlap each other a bevelled fillet must be secured to minimize tripping. Any scaffold exceeding 1.980 m high must be boarded as follows:

men only and no materials	0.640 m wide (three boards)
men with materials	0.870 m wide (four boards)
for supporting higher trestles	1.070 m wide (five boards)
for masonry work	1.300 m wide (six boards)

All platforms exceeding 1.980 m high must be adequately guarded to a height of between 0.920 m and 1.150 m. There should be a toe board, minimum 0.155 m high, or total barier to the outer side and the maximum space between guard and toe board is 0.760 m.

Ladders must comply with BS 1129 Timber Ladders and Steps and should be in good condition at all times. Ladders must be securely fixed to prevent movement (see also page 121).

Where any roof exceeds 10° slope and the surface is slippery or dangerous, crawling boards must be used or a catch barrier erected at eaves level, except where the eaves do not exceed 1.980 m high.

Any loading upon a scaffold must be distributed in a safe way, as near as possible to standards and should not exceed 275 kg/m^2 for general scaffolds.

Part 4: Keeping of records – records of inspections made at seven-day intervals, or after bad weather or after structural change, must be recorded. All records must be retained on site unless work will not exceed six weeks, when they must be retained and be available for the Factory Inspector if requested.

Hoists

Although reference is made later (page 125), a hoist is also an integral part of the scaffold structure. Material hoists are common. Passenger hoists are less frequently used and may create misleading considerations for the reader. Therefore, only material hoists are referred to throughout this chapter.

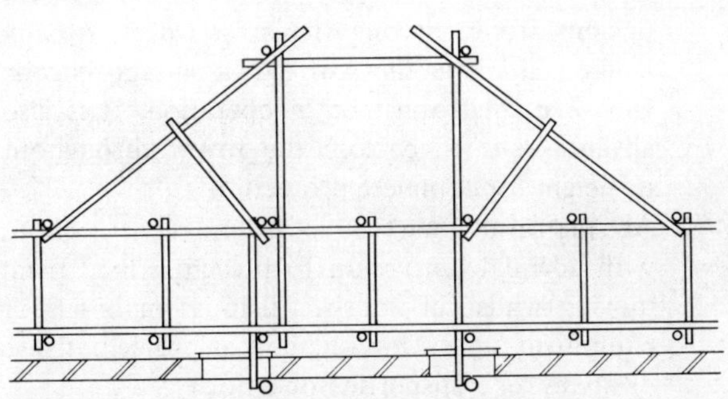

Figure 51 *Hoist tower. This plan shows how bracing and through ties hold the tower into the building to give stability. This is typical of the several different ways this can be done*

The two basic types of hoist are the centre slung and the cantilever. The former is a platform suspended between two masts, which, including the driving apparatus, forms a complete tower. The latter is a platform supported to the side of a mast in a cantilever style. Regardless of type, the erection procedure should follow the following checklist:

1 Adequate support from the scaffold structure and/or the building fabric must be made. The cantilever types need more support because of their smaller construction, so they are supported at approximately 2 m intervals (Figure 51).
2 A hoist tower must be erected plumb to reduce stress to the equipment or interference with platform functions
3 The safe working load (SWL) must be clearly shown
4 A foolproof device to avoid overrun must be fixed to the top of the tower or mast, and be regularly checked
5 The gap between the platform and the landing area to each lift must be as small as practicable and should not exceed 0.051 m
6 The ground supporting the hoist must be firm and, ideally, covered with a concrete base
7 Jacking-up for alignment must be upon firm ground or sole plates
8 The control rope must be operational from only one position and the operator must have a clear unrestricted view
9 At ground level the hoist tower must be enclosed to a height of not less than 2 m to prevent persons from entering the hoist tower. At any landing area, on the scaffold, a suitably strong gate 2 m high must be in operation. It is also advantageous to enclose the tower throughout its height for complete protection
10 All platforms must be of sound construction, with adequate provision to prevent a load from moving whilst in transit. All loose materials or equipment must be retained in wheelbarrows or boxes for transporting on a hoist

For the safe use of a materials hoist there are certain procedures to observe. The main check is for the operator to install the equipment in accordance with the above list. In addition he should observe the following:

1 Reliable training must have been completed before anyone uses the hoist
2 The operator must select a good position for work and arrange trial runs to judge distance. (Good eyesight is an obvious requirement of all transport or machine operators)
3 The operator must familiarize himself with the controls, e.g. the rope which operates movement, the release rope which operates the braking mechanism. Only practice and caution will develop an accurate control
4 Whilst operating the hoist, a safety helmet must be worn
5 All removable, or adjustable parts of the hoist must be checked visually during work, e.g. gates secured in position, load secure, SWL observed

Good maintenance is essential for the safe operation of the hoist. A more detailed control is given on pages 132-3. The following checks should be incorporated with that:

1 A weekly inspection as required by statutory legislation
2 A test and examination before initial use and after any substantial alteration or repair
3 Full examination, with confirmed record, every six months
4 Regular checks to control ropes and operational equipment
5 Checks upon ground buffers, where necessary, and on all devices to avoid overrun

Erecting scaffolds

Much has been written about the erection of scaffolds. The training and certificate issue to scaffolders will do much to improve defects that have occurred in the past.

Putlog and independent scaffolds

1) Prepare a firm base by laying sole plates, large-section timbers not less than 0.225 m x

0.040 m. They are levelled and supported on firm ground. There should be sufficient timber length to support a minimum of two standards to reduce point loading or sinking of individual standards.

2) Standards should be erected plumb with a maximum spacing of 2 m. Each must be held on the sole plate by a base plate nailed or screwed down to the timber. All standards must be parallel to the building structure (independent inner row 0.330 m clearance, putlog 1.270 m from the structure to allow five boards and 0.100 m gap). Joints to extend standards must be staggered to eliminate weakness.

3) Ledgers must be provided at comfortable working heights for the planned job. Joints should be staggered but remain clear of a mid-span position and, preferably, be achieved with sleeve couplings.

4) Transoms used for independent types must be on top of parallel ledgers and spaced according to the boards to be used. There must be a double-up where board ends will meet. Projections beyond the elevation facade of the scaffold should be

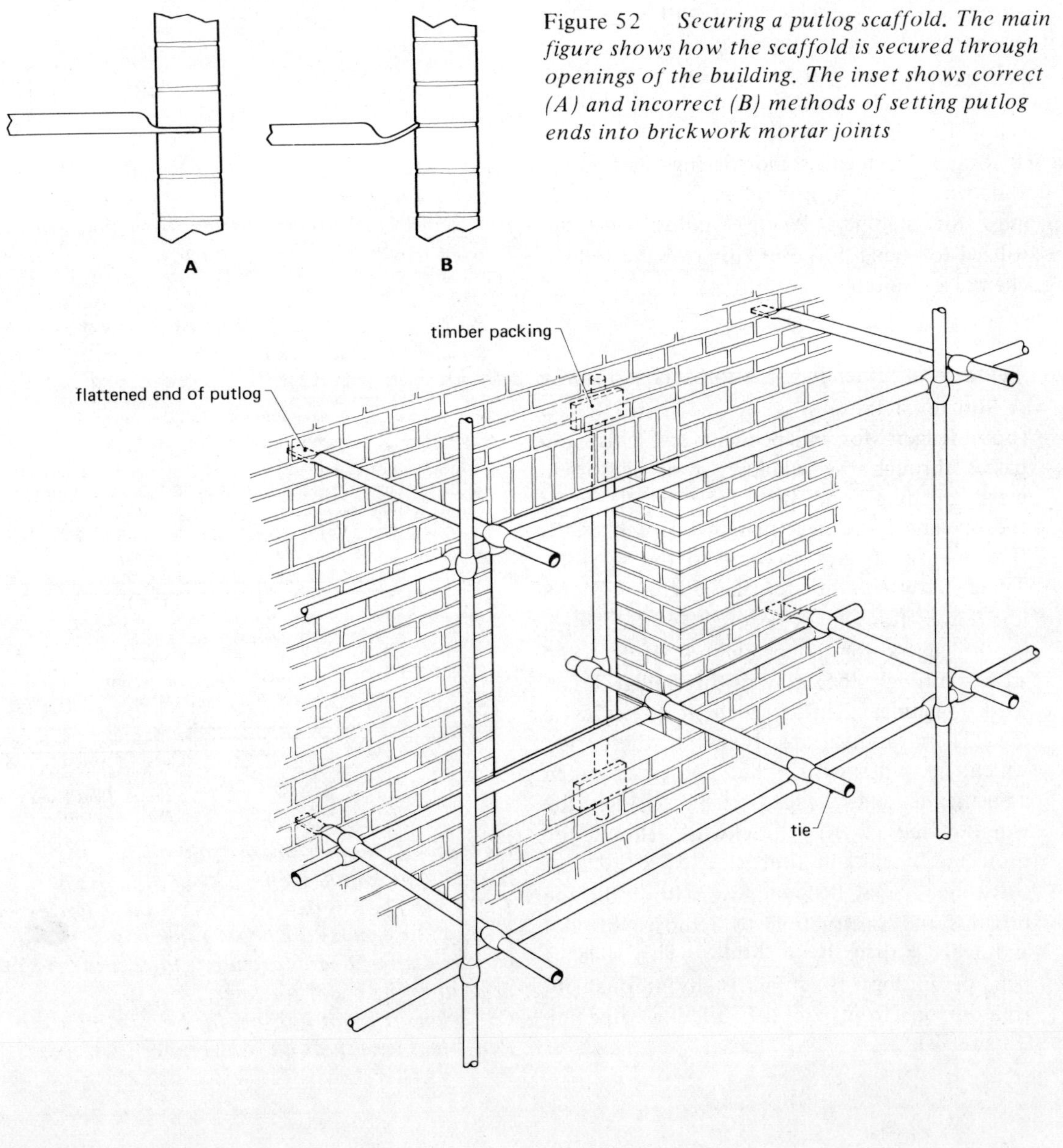

Figure 52 *Securing a putlog scaffold. The main figure shows how the scaffold is secured through openings of the building. The inset shows correct (A) and incorrect (B) methods of setting putlog ends into brickwork mortar joints*

avoided. Where a projection does occur clear warning should be affixed. In situations adjacent to pedestrian areas projecting parts should be painted white and/or sealed with luminous or reflecting torches or a plastic cap.

5) Putlogs for the putlog scaffold must be secured on top of the ledger and secured well into the fabric of the structure. The spade end, to provide improved load-bearing strength, is secured flat into mortar joints (Figure 52). Spacing of the putlogs is dependent upon the boards to be used, and should be 'doubled-up' where board ends meet. Projections of the 'round end' of putlogs is dealt with as described for transoms above.

6) Ties must be secured, ideally to alternate lifts of the scaffold, but at a minimum of 3.900 m vertically. For independent types, the openings to the structure influences and spacings, and putlog types depend upon firm brickwork and suitable openings for stability. Where window openings are utilized for tying purposes either of the following schemes are suitable:

(*a*) Tying with extended transoms (Figure 52). A horizontal transom is secured to the ledger (both ledgers for independent scaffolds) and passes through the window or similar brickwork openings. Another tube which spans the opening by a clear margin on each side is fixed across to give security to the scaffold.

(*b*) Tying between the reveals (Figure 53). A horizontal transom is secured to the ledger (both ledgers for independent scaffolds) and passes through the window or similar brickwork openings. This is secured to a reveal pin. There are various types of these ties, which are tubes with either bolt or threaded adjustment and wedged tight against the window reveal. Any brickwork should be protected by suitable timber, 0.155 x 0.051 m softwood. It may become necessary during the progress of construction to remove the ties and replace them by a 'bridle'. This spans a pair of putlogs to support an intermediate transom in front of the window opening (Figure 52).

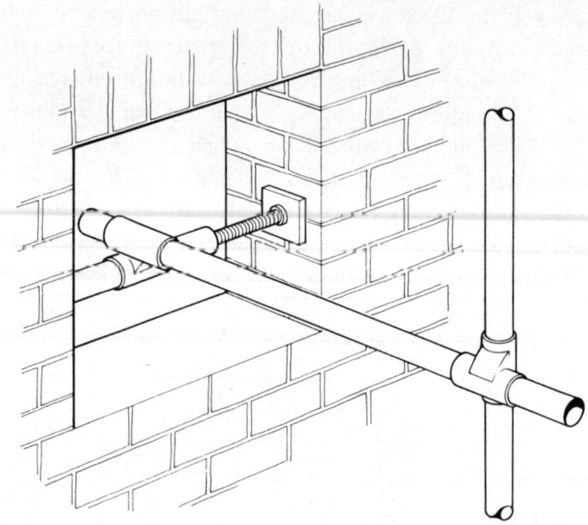

Figure 53 *Tying the scaffold into the building with a reveal pin. The flat plate of the reveal pin is forced out by the thread fitting sleeved into the tube across the window opening*

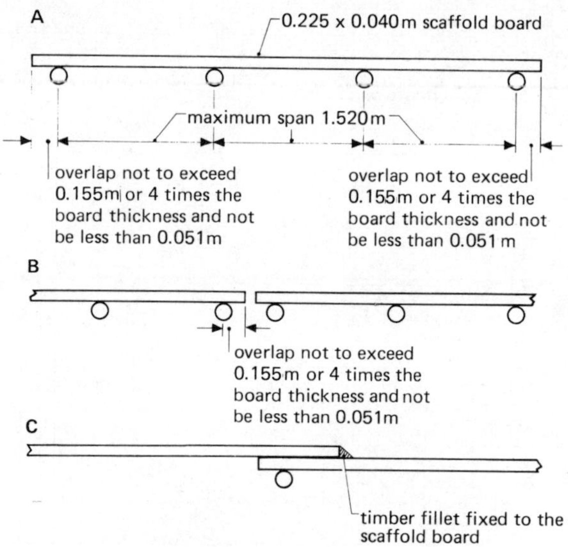

Figure 54 *Supporting scaffold boards.*
A – The maximum span between supports is 1.520 m. The overlap of a board must not exceed 0.155 m or 4 times the board thickness
B – Overlap of boards again not to exceed 0.155 m or 4 times the board thickness
C – Where it is not possible for boards to butt up flat, tripping hazards must be reduced

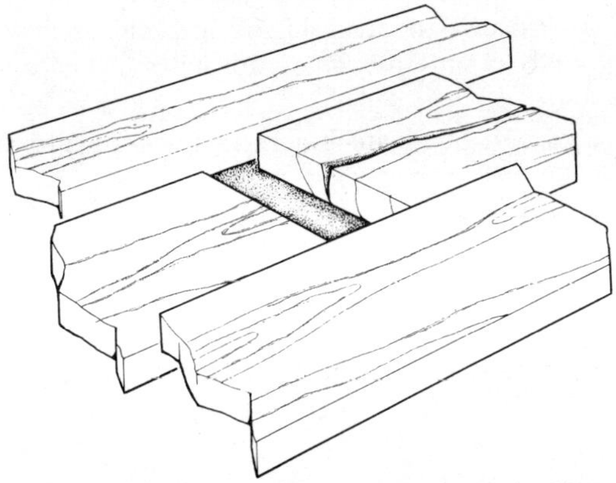

Figure 55 *Dangerous scaffold trap. Anyone using this boarded scaffold area is at risk if he treads near here. All boarded areas must be complete. No board must overlap more than 4 times its thickness. Notice two other faults – one board has an un-tipped end and is badly split*

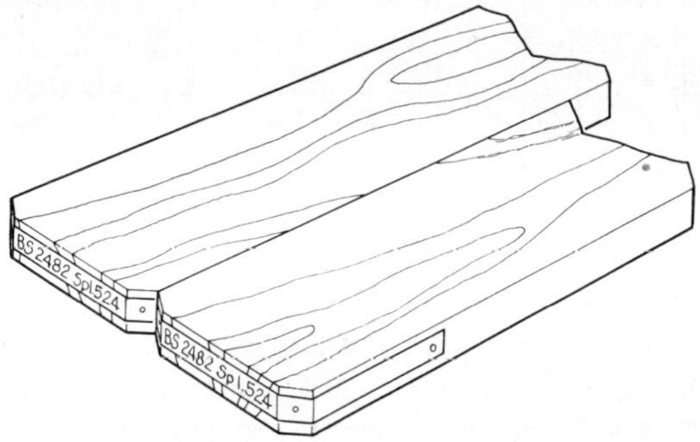

Figure 56 *Tipped ends to scaffold boards. BS 2482 requires all boards to be suitably tipped at both ends to prevent defect or damage. The board is marked on the tipped end with its maximum premissible span and British Standards number, and can usefully be colour-coded (Figure 59)*

7) Bracing must be adequate and secure to both planes of the scaffold. The tubes should be joined with sleeve couplings and fixed with swivel couplings. Either diagonal or zig-zag bracing should be used at the end and front elevations, with fixture as close as possible to ledger fixings.

8) Platforms are prepared from boards which, in accordance with BS 2482 must be of sound structure, free of defects and banded to each end. Also, a clear marking, usually on these metal bands, shows the maximum spans permissible. A minimum of three supports to each board is required and end supports must not allow a board overhang exceeding four times the board thickness (Figure 54). The boards should lay flat; where they overlap, provision to reduce tripping must be provided (Figure 54 (C)). The total platform will vary according to its intended use:

General footways, 0.640 m wide, usually three boards

Walkways with materials, 0.870 m wide, usually four boards

Heavier loads or supporting higher platforms, 1.070 m, usually five boards

Masonry work, including dressing, 1.300 m, usually six boards (only practicable with independent scaffolds)

Masonry work with additional allowance for supporting higher platforms, 1.500 m, usually seven boards (only practicable with independent scaffolds)

The board widths are considered to be 0.210 m. Where a passage for materials is required, a minimum 0.640 m gangway must be maintained. If a direct walkway is required and materials are to be deposited a minimum 0.440 m gangway must be maintained.

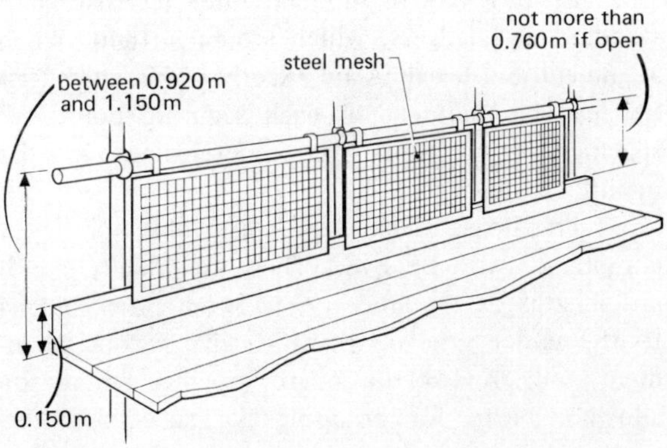

Figure 57 *Complete guarding to the sides of scaffolds. Where work creates risk to people passing beneath – such as above pedestrian walkways, pavements or shopping areas – the scaffold side must have a complete guard to retain materials and debris*

9) Protection must be provided to eliminate the danger of employees, materials or equipment falling, wherever the platform height exceeds 1.980 m. Where the scaffold height exceeds that of the structure, or the gap between building and scaffold exceeds 0.300 m, this provision is needed to all sides; otherwise protection to outer sides and ends is sufficient. Guardrails fixed horizontally between 0.920 m and 1.150 m high are required, along with toe boards not less than 0.155 m wide. The gap between these must not exceed 0.760 m. Where materials are stacked, a complete shield of plywood or mesh is necessary, particularly near pedestrian walkways. Provision for landing, access or egress must be provided correctly, with accurate termination of the guardrail and toe boards. Details of all ladders suitable for this situation and their fixture to the scaffold are shown on pages 118-24.

Interlocking modular scaffolds

Details of assembly vary according to the style of the apparatus. Some types have eye-bolts to tighten with a podger spanner, others require locking bars to be fixed with a hammer. Regardless of this, to achieve a reliable and safe modular scaffold, the following considerations should be taken:

1) Prepare a firm level base with sole plates of large section timbers to receive the adjustable base plates, which in turn will support the standards.

2) Erect a bay of modular units, consisting of standard and ledgers, which are fingertight. When alignment and levelling are exactly correct this first bay can be tightened at each securing point. All spacings of the units are predetermined by the manufacturer's size.

3) Prepare a deck, using scaffold boards in compliance with BS 2482. These are usually standard lengths as required, or, in some cases, blocks to the under side are fixed to give a good alignment and prevent the board moving. Where the allowable span for boarding is exceeded, intermediate supports can be used.

4) Proceed to extend the scaffold horizontally by additional preparation of the ground, sole plates and base plate and erect further base modular units.

5) Proceed to erect higher scaffold as may be required by interlocking more units higher up on the base.

6) Insert adequate braces as the work proceeds and ensure a good zig-zag or cross alignment of braces. Certain types require diagonal braces to retain rigidity.

7) All factors applicable to board decking, toe boards, guardrails, etc., are as those previously referred to under putlog and independent scaffolds.

Tower scaffolds

A static tower scaffold will be supported by the building structure and become a very short version of the independent scaffold unit. The mobile types are prepared in a similar way with a set of lockable caster wheels at the base to give mobility. The main restriction is the ratio of total height to base dimensions. The erection procedure for tower scaffolds is as follows:

1) A firm base must be prepared with sole plates to receive base plates. Where mobile towers are used a firm area of ground must be available for movement, or a clear line of track laid for moving along.

2) Base plates for static, or locked wheel castors for mobile towers should be located at the base of each corner.

3) Upon the base units four vertical corner standards are erected and secured, together with horizontal ledgers and transoms at predetermined lift heights. These should be fixed with right-angled couplings at heights not exceeding 2.750 m or the base size if this is smaller.

4) As the work proceeds, a secure form of bracing must be included, either diagonal or zig-zag style, for the full height of the tower. There should also be diagonal bracing across the corners for additional stability.

5) The tower must be checked for accuracy of plumb, squareness and alignment of all component parts. The termination of towers is restricted to a total height to the platform as follows: maximum of three and a half times the smaller base size for internal use; maximum of three times the smaller base size for external use.

Where these exceed 9.750 m there must be provision to tie the scaffold to the building, or stabilize it by raking supports.

Where the restriction is to be exceeded, a suitable means of additional base structure must be incorporated, to maintain compliance (Figure 44).

6) The working area must be fully boarded with boards in compliance with BS 2482 of minimum thickness 0.038 m.

7) Guardrails and toe boards are needed on every side as detailed for independent scaffolds, between 0.920 m and 1.150 m with 0.155 m toeboards and maximum gap of 0.760 m.

8) Means of access must be provided by vertical ladders secured to the shorter side with a 0.155 m gap above the castors to the bottom edge of the ladder. A minimum of 1.150 m of ladder above the platform must be provided, with a safe means of terminating the guardrail to retain protection but provide access.

Additional points to consider with a mobile tower scaffold are:

Use it only on firm, level ground.
Always clear the platform of materials, tools and personnel before moving it.
Only move it by exerting a force at the very bottom.
Castors must always be locked during use and, preferably, pointed outwards.

Trestle scaffolds

This, the simplest style of scaffold, is recommended only for lightweight work of a short duration. The erection and check procedure is as follows:

1) Always ensure a firm and level base.

2) Inspect the trestles for stability and stand them with stiles fully extended and rungs level, with good alignment. The spacing of trestles is influenced by the boarding to be used. Maximum space between supporting rungs is 1.370 m for 0.040 m boards and 2.440 m for 0.051 m board. Staging boards (Youngman) are superior decking; a minimum of two boards' width must be provided with all lengths equal.

3) No trestle scaffolds are allowed where the user may fall more than 4.570 m to the ground. It is also recommended that the boards should only be placed two-thirds up the height of trestles.

4) An independent means of access, ladder or steps, must be provided where the platform height exceeds 1.980 m. Guardrails and toe boards are not usually required where a fixed trestle, with two tiers of scaffold, is used unless the platform height exceeds 1.980 m.

Work on roof areas

Where work is to be done on a pitched roof, special provisions are needed to reduce risk of falling. Pages 118-24 deal with all roof works from ladders. Also, at the edge of a pitched roof something is required to prevent a fall directly to the ground. The usual procedure it to use directly, or adapt, the top platform of the general scaffolding (page 107).

Dismantling and maintenance of scaffolds

It is a good policy to complete all maintenance of scaffolding during dismantling. If all damaged pieces are discarded or repaired immediately, there should be no risk of their re-use. The following rules apply during dismantling:

1) An organized, systematic approach is required that does not allow collapse by irresponsible dismantling. Where temporary bracing or support is needed it must be incorporated, as must all warning notices and barriers at the end of each working shift or day.

2) Every scaffold must be inspected after erection, at every seven days thereafter, after structural alteration or after bad weather. The details must be recorded within Form 91 Part 1 Section A (pages 38-40).

3) Where wind might create a hazardous situation, a suitable means of holding down all loose boards must be provided and implemented.

4) It is good practice to turn back the boards adjacent to the structure to avoid dirty splashmarks. It is a good safety procedure to turn over all scaffold boards immediately after snow or frost, to reduce the risk of slipping on dangerous boards. The alternative is to use sand on snow but sand may blow about afterwards.

Scaffold Safety Checklist for use at Inspections

Description and location of scaffold

Type of scaffold	Putlog Independent tied Special	**Permissible load per square metre**		**Hoist tower** Yes/No
Materials used:	Galvanised	Black	Alloy	Other

At each inspection check that your scaffolding **does not** have the faults described below.

Footings	Soft and uneven	No base plates	No sole boards	Undermined
Standards	Not plumb	Joined at same height	Wrong spacing	Damaged
Ledgers	Not level	Joint in same bays	Loose	Damaged
Putlogs and transoms	Wrongly spaced	Loose	Wrongly supported	
Couplings	Wrong fitting	Loose	Damaged	No check couplers
Bridles	Wrong spacing	Wrong couplings	Weak support	
Bracing: facade	Some missing	Loose	Wrong fittings	
Bracing: Ledger at right angles to building	Some missing	Loose	Wrong fittings	
Ties	Some missing	Loose	Physical not enough	Reveal not enough
Boarding	Bad boards	Trap boards	Incomplete	Not enough supports
Platform	Not wide enough			
Loading	Too heavy	Shuttering propped from scaffold		
Brick guards	None			
Guard rails	Wrong height	Loose	Some missing	Wrongly positioned
Toe-boards	Wrong height	Loose	Some missing	
Ladders	Damaged	Insufficient length	Not tied	
Access	Obstructed	Not enough		
Gin wheels	Weak supports	No identity number	Hook not moused	No check fittings
Fans	Weak supports	Not enough guy wires	Some missing boards	No hand rails
Hoist towers	Not enough ties	Not enough fencing	No gates	Poor operating position
Mobile Tower Scaffolds				
Height	Too short			
Height ratio Base	Indoor, more than 3½ to 1		Outdoor, more than 3 to 1	
Ties guys or base weights if needed	Some missing			
Surface	Soft		Uneven	Sloping
Access	Ladder not provided		Insufficient length	Ladder not tied
Bracing	Some missing		Wrong direction	
Brakes or chocks	Not provided		Not secured	
Wheels	Liable to fall out			
Guard rails	Wrong height		Loose	Missing
Toe-boards	Wrong height		Loose	Missing
Temporary Roofs and Beamed Scaffolds				
Design drawings	Not provided		Not sufficiently detailed	
Scaffold	Not in accordance with drawings			

FOR GUIDANCE ON GOOD STANDARDS SEE BS CP 9.7.

** A recommendation of a Sub-Committee of the Joint Advisory Committee on Safety and Health in the Construction Industries.*

Figure 58 *Scaffold safety checklist for use at inspections*

5) Keep an even distribution of loading and avoid point loading at any particular place.

6) Retain a clean scaffold whilst in use and make special checks for slippery boards after bad weather.

7) Clean, check and maintain all scaffold parts as they are dismantled. Any damaged or otherwise unusable items must be repaired or discarded.

Storage of scaffold components

The correct storage of scaffold components indirectly affect future safety and keeps these expensive parts in good condition for as long as possible.

Scaffold tube members

These must be stacked flat, preferably off the ground and in a suitable rack. A further advancement is to colour-code the tubes in order of length, e.g. 3 m length – red, 3.5 m length – blue, etc. A simple band at the end 0.155 m wide would be sufficient to advise the user, and simplify selection.

Scaffold boards

A racking arrangement is satisfactory for small amounts; a large quantity, however, need to be stacked off the ground. It is good practice to colour-code the boards also, and to store boards in 'sets' of five where used in general scaffolding. If boards are to be stored for a long time, or have been soaked prior to storage, it helps to 'batten' the pile, as with timber seasoning, to allow through the stack a clear passage of air, which improves drying out and reduces fungal attack.

Fittings

All similar fittings must be kept together and separated from other types. As fittings are stored a check must be made and all moving parts oiled for easier use.

Timbers

All timbers must be stacked flat to reduce twist and all nails or similar projecting objects removed.

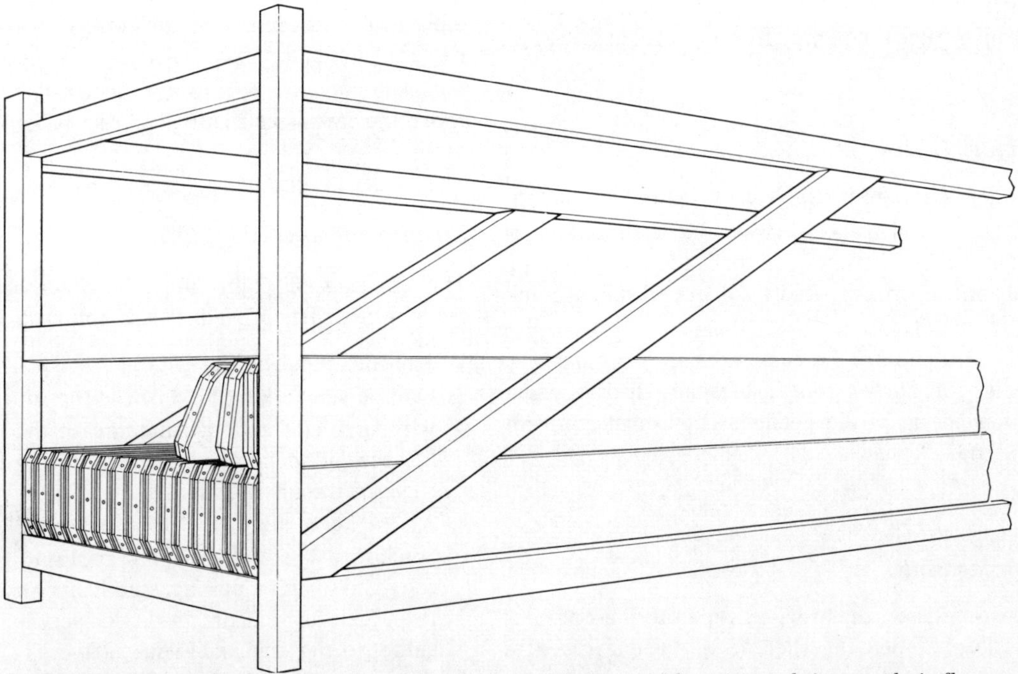

Figure 59 *Scaffold boards. Boards should be stacked on edge to avoid moisture lying on their flat surfaces. A simple colour code marked on the metal-tipped end makes finding the right board easy: red – 3.000 m, blue – 3.500 m, yellow – 4.000 m*

12 Ladders

The function of a ladder is to allow ascent to or descent from a working place above or below normal ground level. In general terms the requirements are similar for working up and down ladders, to higher or lower work positions. The legal controls upon ladders are detailed within the Construction (Working Places) Regulations 1966 reviewed on pages 107 and 176. Ladders can be a means of access to a scaffold, or to an individual working place.

A ladder, irrespective of its style, is simple and versatile. There are certain situations within the completed building where permanent ladders are required, e.g. as roof access for maintenance. This chapter is only involved with temporary ladders of which the following are the principle types.

Plain rung ladder

This type (sometimes referred to as pole ladder) is used for general work processes and is described by the number of its rungs. Timber ladders are economical, but are heavy, easily subject to damage and require maintenance. They must conform to BS 1129. Aluminium ladders are more expensive initially but have a longer lifespan, during which maintenance is hardly needed. They must conform to BS 2037.

Extension ladder

This is similar to the previous type but incorporates two pieces joined together, to allow one piece to slide over the other and be retained by a hook over the lower rungs. The ladder can be extended or retracted by pulley ropes.

Step ladder

This is a free-standing unit with two sides separated to give stability. The two parts are hinged, and strong cord prevents the legs from spreading. These types of ladder are controlled by the same British Standard requirement as those for plain rung ladders, BS 1129 for timber and BS 2037 for aluminium.

Roof ladder

This is an adapted rung ladder with a hook on the top end for securing over the ridge of a roof. They are used in conjunction with plain rung or extension ladders and are always necessary for access above the eaves (see Figure 49, page 108).

Erecting and lowering ladders

The erection of plain rung and extension ladders are identical. The procedure is as follows:

1 Lay the ladder on the ground
2 One person, taking hold of the top rung with the wire or metal support underneath the rung, can lift the ladder and commence moving his hands towards the other end
3 At the same time, the second person, acting as 'anchor man' stands with one foot on the bottom rung and exerts a pulling, stabilizing effort
4 In a careful unison the two people pull the ladder to the vertical (Figure 60)
5 One can now transfer the ladder to its required position using the lifting action shown in Figure 61. It is important that a firm, bracing hold is

Figure 60 *The correct way for two people to raise a ladder*

used with the ladder securely held into the carrier's shoulder

6 Place the ladder against the building or scaffold at an angle of 75 degrees, that is a vertical-to-horizontal ratio of 4:1 (Figures 62, 63 and 64)

Where a short ladder is to be used, one person can erect the rung ladder as follows:

1 Place the bottom of the ladder firmly against the base of the building or a similar fixture
2 Lift the top of the ladder, and push upwards to raise the ladder to a vertical position
3 When the ladder is vertical, transfer it to its required position as described above and shown in Figure 61 (B)

Step ladders are mainly reserved for interior work; the erection procedure is:

1 Stand the ladder on a firm level area, or prepare a suitable platform with scaffold boards
2 Position the legs as far apart as the retaining cord allows
3 Check the step ladder is level before ascending

Securing of step ladders is not usually required.

Where roof ladders are to be employed, a plain rung ladder or extension ladder must first be erected as previously described, and secured as explained previously. The procedure of erecting a roof ladder is as follows:

1 Proceed up the lower ladder to eaves level, with the roof ladder being carried up or passed by an assistant

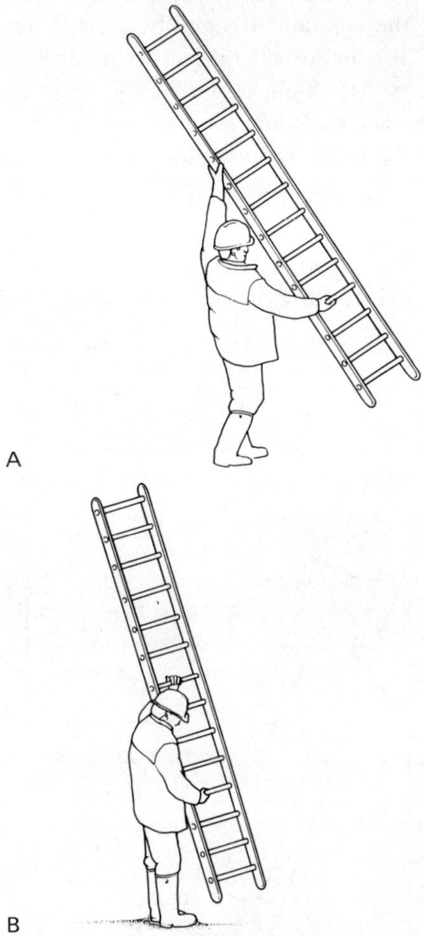

Figure 61 *Transferring a ladder single-handed. A – Wrong stance. The operator has little control and an accident may easily occur B – Right stance. With this grip the operator has much better control of the ladder*

2 Pass the roof ladder up the slope of the roof with the anchor hook turned uppermost
3 When the anchor hook has passed the ridge, turn the ladder over and lower the roof ladder to engage the anchorage. Also align the ladder stiles
4 Protect the roof weathering, i.e. slates or tiles, with a straw-stuffed hessian bag placed under the roof ladder
5 Secure the access by lashing the ladders together with rope

The method of lowering any ladder is a reverse procedure of the erection. It must be carried out with caution. It is important that all fixing devices to give security are retained until they *must* be removed and also that the lowering of ladders, particularly long ones, is done by two persons.

Securing ladders

Ladder access is a temporary arrangement, but must be prepared and fixed in a secure manner. The fixing of a step ladder and roof ladders has been mentioned. For plain rung and extending ladders follow these rules before securing the ladder:

1 The angle at which the ladder leans against the building must be 75 degrees or a height-to-space ratio of 4:1
2 All ladders must extend at least 1.066 m above the landing or above the rung height a person needs to stand at. This allows a good hold for moving from the ladder to the scaffold or working from the ladder. (See Figures 62 and 63)
3 Ladders should not project by an unnecessary margin beyond the platform
4 The ladder must be inspected prior to installation. Splits, cracks, worn rungs, loose rungs or any similar defect must be properly repaired or the ladder discarded

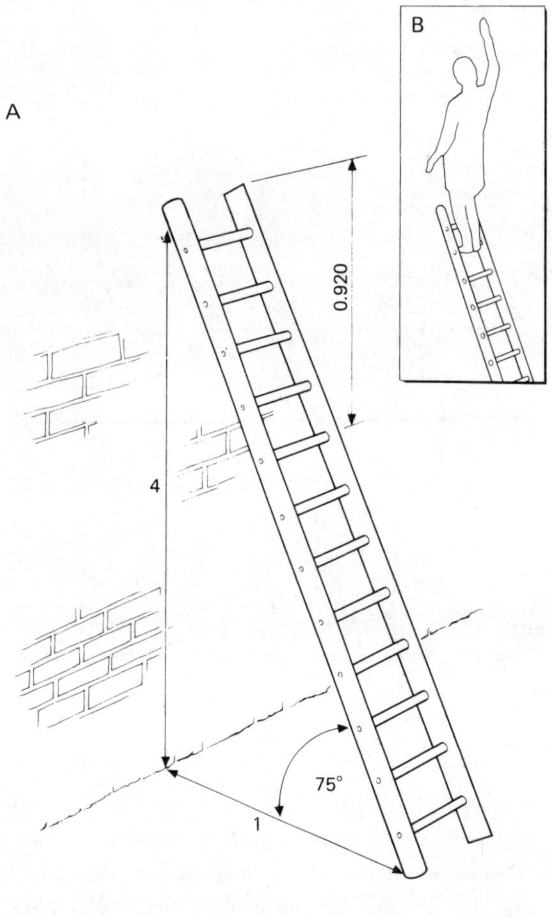

Figure 62 *Allow an adequate length of ladder!*
A – The correct angle for the ladder is 75°; this is 1 m out for every 4 m of vertical height
B – Over-reaching is dangerous. A good allowance of ladder – at least 0.920 m – above the employee's feet is vital to ensure a grip in the event of slipping

All ladder access must be secured to prevent any slipping or displacement that will cause danger. The following systems of securing a ladder are acceptable individually or in combination:

1 Tie the higher part of ladder stiles to the ledger of scaffolding structure and maintain a clear footing to the rung as well as 1.150 m extension above the platform (Figure 63)
2 Drive in a peg, or pegs, adjacent to the bottom of the stiles and tie these stiles to the pegs (Figure 64). This can only be done on open ground
3 Secure the higher part of the ladder stiles to a fixed anchorage of the structure (Figure 64). It is good architectural and building practice to provide such anchorage permanently at the design and/or building stage in certain building styles
4 Where the work is of short duration but on a hard surface, e.g. window-cleaning, rubber pads are available for the base of stiles
5 Where work is of short duration and no form of fixing can be achieved, someone must 'foot' the ladder to hold it firm against movement
6 Where the ladder is to be used at a constant height, e.g. within a store, hooks to the top of the ladder can be engaged on to a continuous tubular rail (Figure 65, page 122)

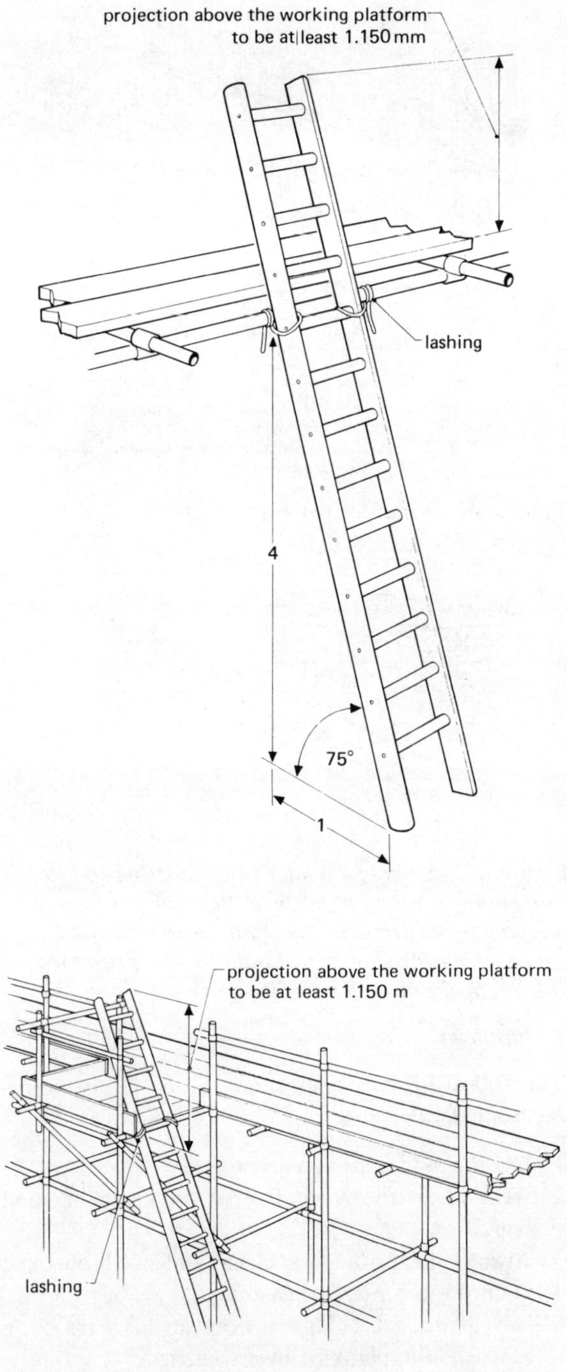

Figure 63 *Securing a ladder to a scaffold. The ladder must project at least 1.150 m above the platform and be leashed securely. The angle is the usual 75°*

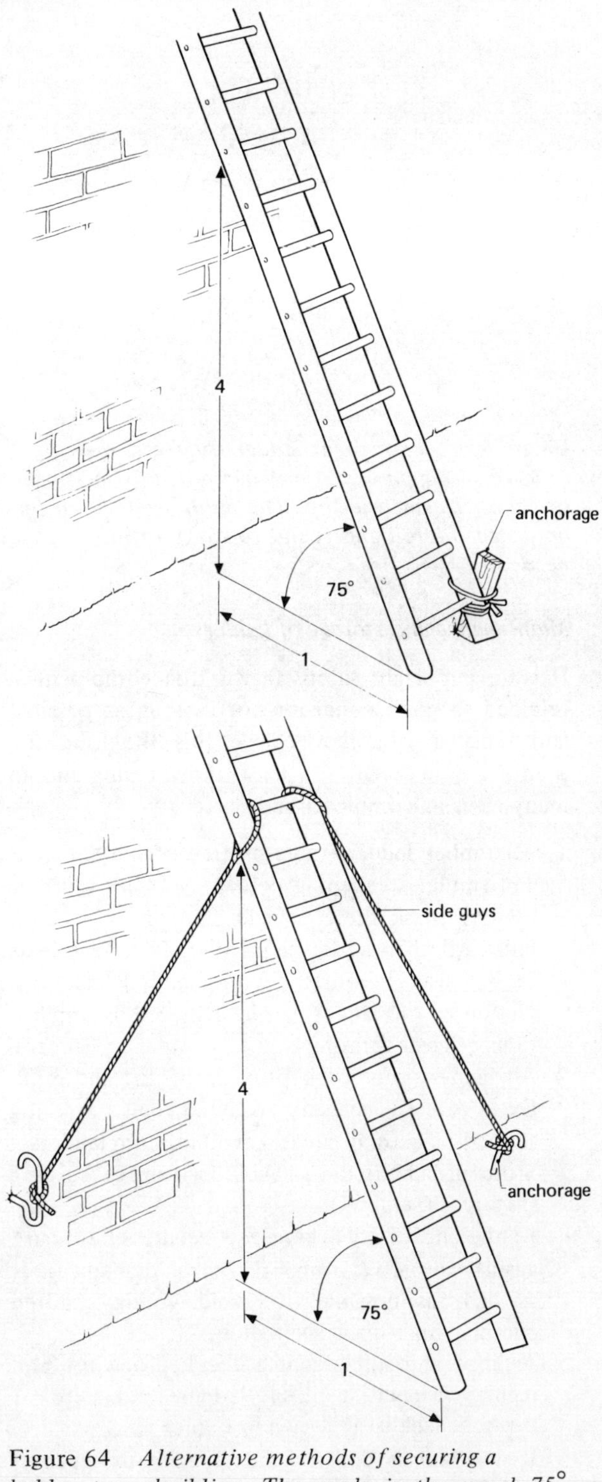

Figure 64 *Alternative methods of securing a ladder to a building. The angle is the usual 75°*

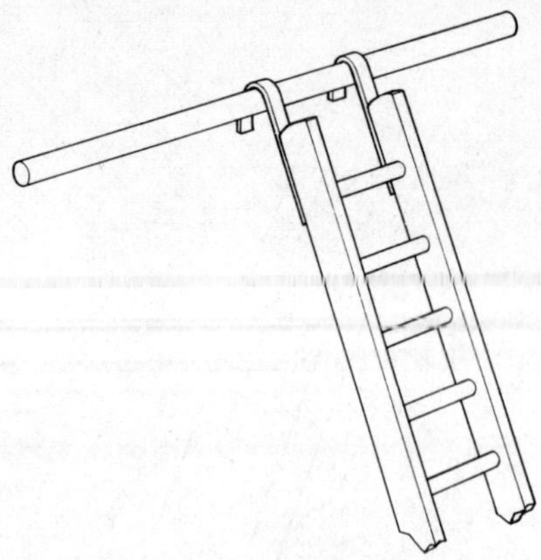

Figure 65 *Permanent ladder for fixed shelving in large storage areas. A tubular rail runs in front of the shelves at the top. The ladder is secured by hooks to the rail and is slid to the position it is needed*

Maintenance and storage of ladders

It is an important safety factor that equipment is retained in good condition for as long as possible and remains safe. To achieve this the following points are necessary at least individually and in many cases in combination:

1 All timber ladders must be free of splits, cracks and similar weakening defect. No filler must be used to mask these defects, and on no account must a timber ladder be painted. To hide defects in this way is morally wrong and illegal. Protection of the ladder can be achieved by applications of clear varnish
2 All ladders must be kept clean of dirt and mud. Extra care is necessary to avoid their becoming slippery, e.g. oil or grease on aluminium ladders
3 Store all ladders flat to avoid twisting or warping (Figures 66 and 67)
4 At the end of a working day, secure a flat board against the lower rungs of the ladder and tie it as high as possible, to avoid young children climbing up after working hours
5 Arrange and implement a check procedure and regular inspection of all ladders and steps. A typical checklist is shown in Figure 68
6 If a ladder is defective, destroy it or repair it immediately before anyone can use it by mistake

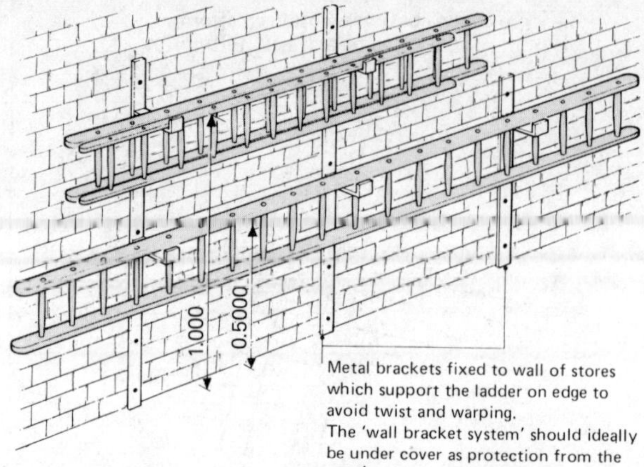

Figure 66 *Ladder storage in a rack*

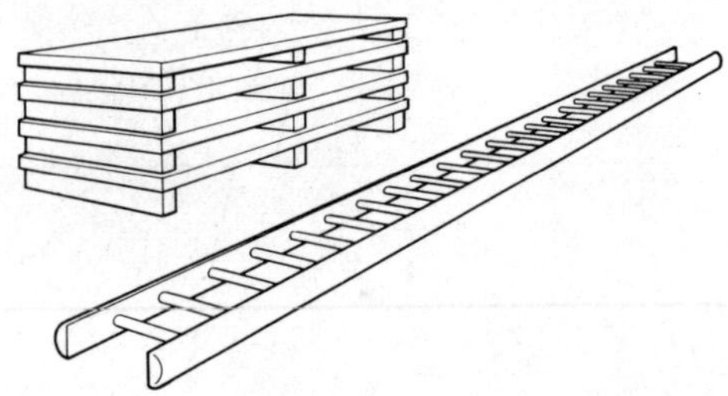

Figure 67 *Storing a ladder for a short period. The ladder is on a flat surface that will not cause warping or twisting. If the ladder is to be stored for any length of time, it should be in a protected store – again on a flat surface*

In summary

The following is a brief summary of safety points regarding ladders:

1 Use the ladder at its correct angle
2 Wear clean footwear, free of mud and in good condition
3 Always keep ladders clean and well checked
4 Do not reach out from a ladder
5 Never support a ladder from an insecure base, e.g. scaffold, planking over trenches
6 Always secure the ladder or get someone to hold it
7 Wherever possible, carry tools in pouches or belts or have them hoisted to the work place

R & D Building Company

Pararad Road, Roselip

SCHEDULE OF LADDERS

LADDER REFERENCE	1978 ACTION	1979 ACTION	1980 ACTION	1981 ACTION
LADDER A 1				
LADDER A 2				
LADDER A 3				
LADDER A 4				
LADDER B 1				
LADDER B 2				
LADDER B 3				
LADDER C 1				
LADDER C 2				
LADDER C 3				
LADDER C 4				
TRESTLE 1				
TRESTLE 2				
TRESTLE 3				
TRESTLE 4				
STEPS 1				
STEPS 2				
STEPS 3				

Figure 68 *Schedule of ladders. Reference numbers indicate different styles and lengths of ladder. Each piece of apparatus is marked with its own number. The columns are filled as repairs and checks are made*

8 Store ladders to retain their good condition
9 Only use firm, steady, authorized ladders
10 If unsure in any way keep off the ladder

The following is a checklist for the inspection of a ladder:

1 The ladder must be generally stable and rigid
2 All parts of the stiles must be free from defects, e.g. splits, cracks or timber decay
3 All rungs must be firm and free from defect
4 All rung supports must be in position and securely fixed
5 All fittings and appliances, e.g. ropes, pullies and hooks to extension ladders, must be in good working order
6 All rungs must be securely wedged
7 Clear varnish preservation must be applied as necessary

13 Site transport

The mode of conveyance of materials and manpower around the construction site is 'site transport'. This should not be confused with 'site plant', which is a broad description of the machines or apparatus actually used for construction work. Mechanical plant is the machines used for site movement, e.g. excavators. Non-mechanical plant is static equipment used for construction work, e.g. scaffold, ladders, trench supports. Site transport is therefore those mechanically propelled machines or vehicles engaged in movement of materials and manpower, e.g. dumpers, hoists, etc. This chapter deals with the general concept of site transport, without detailed reference to any particular machine.

Human error

Wherever people do a job, there exists a possibility that their activities may be defective, inaccurate or incorrect. This may be through lack of knowledge, lack of ability or simple error.

Loading

This can be incorrectly done through lack of consideration for the machine. It is often not realized how deceptive the ratio of weight to volume can be, and misjudgement of the load can follow.

Overloading causes excessive pressure upon the sprung parts of the machine, including the chassis or body frame. The mechanical transmission is also strained, with a subsequent reduction in efficiency and life span of the engine. In addition to the strain inflicted, the machine becomes dangerous at cornering and less effective when braking. The obvious dangers are those of crashing or overturning of the machine with the subsequent injury or death of those involved.

Hoists that are overloaded may crash, overturn or possibly stall midway through transit. A possible remedy is to incorporate a warning light or audible warning that operates when the maximum loading of the machine is reached. The 'safe working load' (SWL) can be marked on to the machine, but will prove to be of little advantage without knowledge of the materials or manpower weight to be loaded. Therefore, an automatic warning is needed to indicate when SWL is exceeded.

Poor load distribution – where the load is distributed unevenly the frame or chassis becomes unbalanced and strained. The machine is therefore dangerous at corners or during braking, and creates serious dangers to employees.

Insecure loading – regardless of its size, shape or weight, the load must always be affixed to the machine to eliminate any danger of movement. Loads that are insecurely fixed create manoeuvring difficulties, braking becomes inconsistent and the chances of overturning are greater. Also, the load may move or become unbalanced on the machine and can slip off its loaded position.

Projected loads – where the size of load exceeds that of the machine, the resultant overhang creates safety hazards. Any projecting structures are vulnerable to damage and may endanger the machine operator or others in the vicinity. Where an overhang is unavoidable there must be clear warning by lighting or bright cloth material. Where necessary, the large load can be accompanied by another machine and/or attending employee on foot.

Drivers – all drivers of vehicles must, by law, leave their cab during loading by crane or similar loading device. It is also good safety practice to check the vehicle at completion of loading and prior to moving off.

Starting

The method used to activate a machine varies according to its source of energy or power. Regardless of size, type or use of machine, anyone who starts it must be advised, and where necessary trained, in starting procedures.

Starting by handle – this activates a mechanical engine by turning it over with a starting handle until the engine is sufficiently excited to power itself. Care must be taken to avoid physical damage to the hand or wrist should the engine backfire.

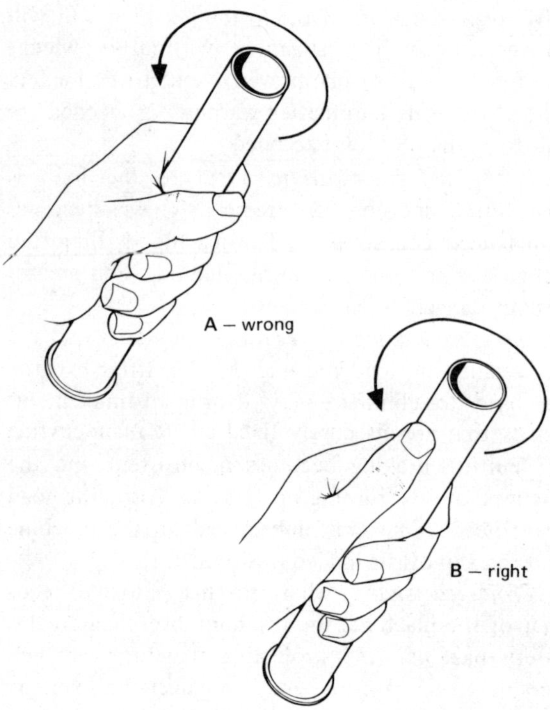

Figure 69 *Avoid accidents with starter handles. A – Wrong grip. The hand is completely round the handle. Kickback (indicated by the arrow) will force the handle against the thumb, causing injury B – Right grip. The thumb is along the shaft of the handle. Kickback will throw the handle harmlessly out of the hand*

Should this happen, the handle jerks backwards against the thrust of the operator and may break thumb, wrist or forearm. Kickback cannot be avoided or anticipated. To avoid injury, the handle must be held so that the thumb is not wrapped around it, but aligned with the fingers (Figure 69). This will allow complete freedom if the handle kicks back.

Electric starting – a sealed circuit should render the starting procedure almost harmless. In the event of failure the safety checks and corrective measures, e.g. fuses, connections, etc., must be completed as described on pages 134-41.

Setting off

The initial moves of any mode of transport require care, caution and consideration. Irrespective of the direction of movement, the operator must check all around the machine.

In the case of a horizontal manoeuvre the driver or operator must check:

1 *Behind*, to avoid crushing persons or objects
2 *To the sides,* to prevent crushing persons or objects
3 *Above*, to avoid projections, cables, etc.
4 *All around*, to avoid excavations

In the case of a vertical manoeuvre the operator must check:

1 *Above*, for clearance of travel
2 *Below*, to ensure safety rails and gates are positioned
3 *All around*, to avoid any projections or cables

Where the driver has not a clear, unrestricted view he must have the help of an assistant to ensure all dangers are eliminated. A clear and concise common procedure must be maintained for all advisory movements. Such communication could be achieved by:

1 Visual hand movements
2 Direct verbal instruction
3 Indirect verbal instruction by intercom/short wave radio
4 Warning lights, e.g. reversing lights

R & D Building Company

Pararad Road, Roselip

PLANT DEPARTMENT

COMPETENCE CARD

Name ....

The company herewith confirms the above named employee to be trained and competent to drive on construction sites, the following plant or vehicles.

1. ....
2. ....
3. ....
4. ....

In addition the same .... has completed and passed the company standard test for such plant on ....
and is an authorised driver until ....
NB. These conditions may be repealed at the discretion of the company Safety Officer subject to any incident that may occur.

Signed ....

Figure 70 *Competence card for vehicle driver*

R & D Building Company

Pararad Road, Roselip

PLANT DEPARTMENT

DRIVERS ADVISORY CARD

The following must be observed at all times.

1. Keep machine in good clean condition.
2. Check and preserve tyre pressure, oil level and fuel level.
3. Have no naked flame when refuelling.
4. Report all defects immediately.
5. Carry out periodic checks as necessary for type of plant.
6. Maintain a clear and concise logbook.
7. Check and maintain guards to all moving parts.
8. Start the machine in accordance with training scheme.
9. Maintain good loading of plant.
10. Check, maintain and drive the plant as instructed and to the best of your ability.
11. Never abuse or use recklessly any plant or machine.
12. Do not drive a poorly loaded machine.
13. Do not check or adjust any machine when it is running.
14. Never carry passengers unless using suitable plant.
15. Do not leave a machine running whilst stationary
16. Never remain at your seat during loading.
17. Do not improvise where repair is needed.
18. Never commence travel without good checking around the plant.
19. Do not exceed acceptable speed limits.
20. Do not ignore any training advice or do anything that may be dangerous or hazardous.

Figure 71 *Drivers' advisory card reminds drivers of safety precautions*

Crane drivers particularly need assistance, and special hand signals are applicable to site transport movement of all machines. All drivers, operators and banksmen should be trained and conversant with a clearly understood system of signalling.

Driving faults

The subject of most criticism is the poor standard of site driving. The lack of ability of the driver and his ardent, sometimes ruthless abuse of the equipment cause much concern and create serious risks. The Construction (General Provisions) Regulations 1961 require that all drivers must be trained and competent. The control of this statutory requirement is often inadequate, and the following should be considered to remedy the problems of bad site driving:

1 A good practice of the employer is to check the driving licence of all potential drivers. Endorsements give proof of bad driving
2 All drivers must be eighteen, unless under training to become a site driver
3 Qualified and competent drivers should be issued with a competence card, to indicate what he is able to drive. These should be renewable periodically, say every three years (Figure 70)
4 Training is needed to provide initial knowledge and to refresh experienced drivers. This can serve as guidance for the issue of competence cards. The training must be carried out on a special two-seat training vehicle either by the company safety officer, plant manager or at a training establishment
5 The company must maintain a list of drivers qualified for site driving
6 A bonus or incentive scheme could be implemented to encourage drivers to maintain a good safety standard

Notwithstanding all the advisory, compulsory or company standards there are moral and legal obligations on all employees to work safely and this includes careful site driving.

Passengers

It is an offence to travel on construction site plant unless it is specifically designed and made for that purpose. Riding on or being lifted by any part of a machine is dangerous. The employer must provide adequate passenger transport – vehicles with raised sides, having fixed seating and passenger hoists. The employee must not risk his safety so must use only the permitted mode of transport. Clear notices must be displayed to ensure everyone knows how to comply with the law, e.g., 'No riding on this vehicle,' 'Keep off – no passengers,' 'Carrying of passengers is illegal on this machine.'

Falls

Whenever plant slips into an uncovered inspection chamber or a large excavation, there may be serious injury and possibly death. To avoid such unnecessary accidents, the following precautions should be observed, individually or in combination:

Fence off hazards

All hazards that may endanger the site transport must be marked, and a clear margin allowed for unstable soil adjacent to the excavation. Small holes, inspection chambers, etc., must be covered with a sheet of metal or similar material with an ample reserve of strength (Figure 72). Trenches and similar excavations need to be surrounded by substantial fencing of timber planks located into slotted, free-

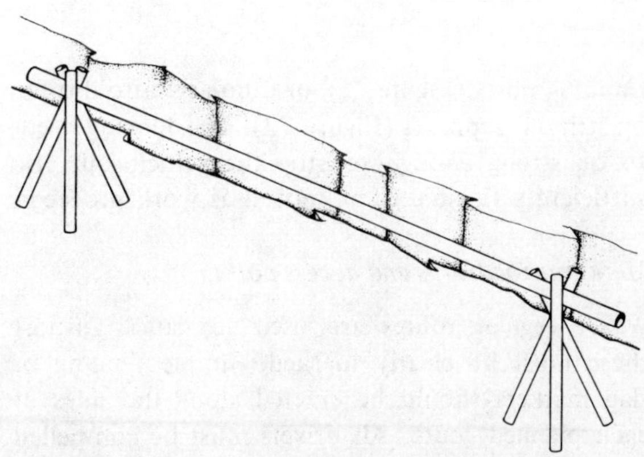

Figure 72 *Fencing off trenches with scaffold tubes. Tripods of short lengths of tube are set at regular intervals. They support longer lengths of tube which act as horizontal barriers*

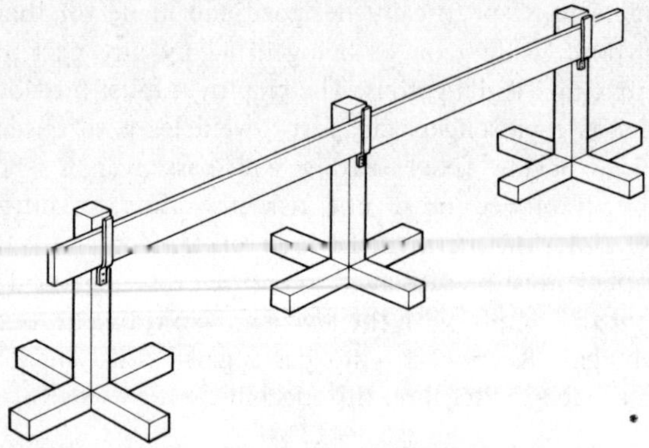

Figure 73 *Fencing off trenches with timber frames and rails. Small frames are made up with short lengths of timber – 0.100 x 0.051 m – using a simple joint base to support the vertical post. Horizontal rails are fixed to the post with nails or mild steel brackets*

Figure 74 *Covering holes. Small holes and excavations can be covered overnight or for short periods with timber or sheet metal*

standing posts (Figure 73), or tubular scaffold tubes on 'tripod' supports (Figure 72). The fencing needs to be strong enough to stop the vehicle but also sufficiently flexible to be moved as work proceeds.

Mark out roadways and access points

Where regular routes are used for site transport these must be clearly marked. Simple fencing or flag markers should be erected along the sides of each planned route. All drivers must be compelled to keep within the confines of these boundary marks. All marked routes must be clear of any obstruction or excavation that may cause danger or hazard.

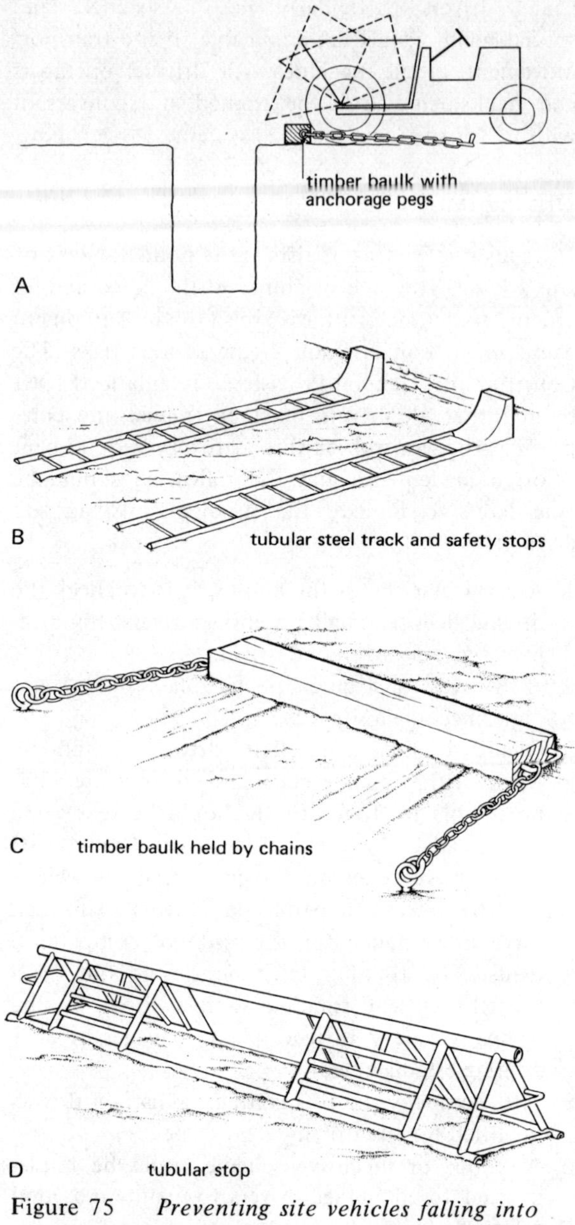

Figure 75 *Preventing site vehicles falling into an excavation when tipping.*
A – Large baulk of timber (0.225 x 0.225 m) is secured to anchorage pegs that are well clear of the ground affected by the excavation (i.e. clear of the angle of repose)
Alternatives are:
B – Tubular steel track and safety stops
C – Timber baulk held by chains
D – Tubular stop

Attendance and guidance

Where the driver's view may be restricted, an attendant should be employed to guide the machine away from ground hazards. This is unrealistic and uneconomical for several trips, but is sensible for one-off transport jobs.

Preventive barriers

A frequent cause of accident, with site transport, is falling into an excavation during discharge of load. Usually the action is one of reversing and the driver may not have a clear view. A large baulk of timber should be secured along the sides of the trench or excavation, against which the vehicle can gently bump its rear wheels (Figure 75). A good allowance must be provided to avoid the collapse of trench sides, or dislodgment of adjacent materials.

Collisions

Operatives

Operatives' work-place should be carefully planned to avoid any chance of collision with site transport. If work on or near to a site roadway must be carried out, adequate warning notices and protective barriers must be erected. Where space does not permit, this work must be undertaken outside normal working hours, or the roadway temporarily closed. Men working on the site must show respect for site transport and keep clear whilst machines are moving, especially where the driver's view is restricted.

The structure

Damage to the completed or part-completed building should be avoided, not only in terms of monetary or time loss but as a safety factor. Barriers or painted drums must be erected around vulnerable areas of the structure. Where transport is required to pass between fixed structures, temporary barriers should be erected which act as guides to show the width through which the vehicle can safely pass.

Materials

To avoid collisions with stacked materials, it is important to plan all storage clear of site roadways. Economical off-loading of bricks, timber, etc., is achieved by stacking directly from the lorry to the space adjacent to the site road. This is acceptable providing moving site transport is not endangered or restricted. Temporary barriers, warning lights or bunting flags can easily be used to mark the area around stacked materials and guide machine drivers well clear of the hazard.

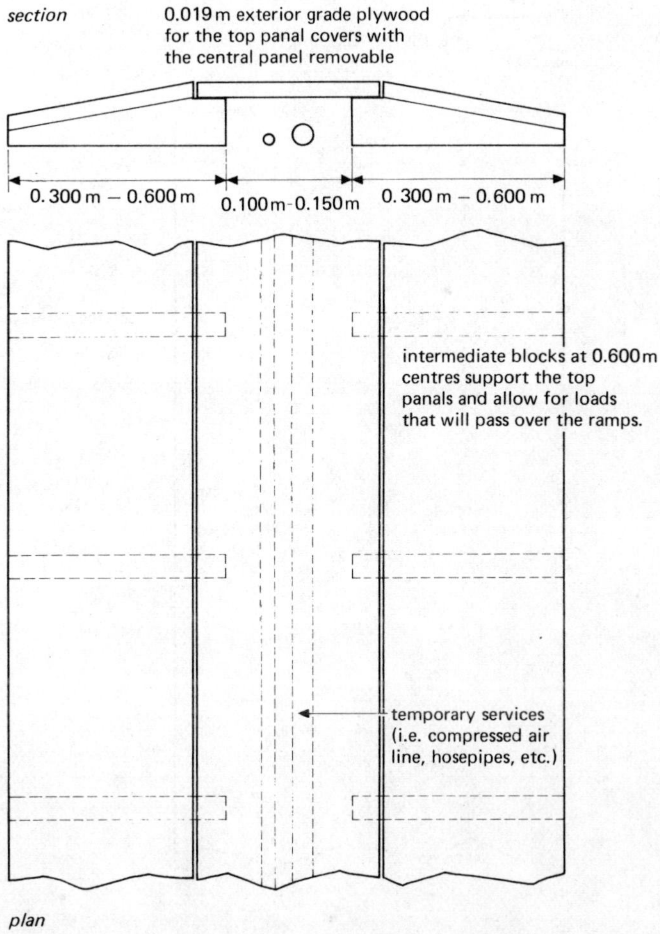

Figure 76 *Ramp to protect temporary services on site. The expense of building ramps is avoided on most sites. But the cost is less than it appears as reuse is possible, and lives are not endangered by damaged site services*

Obstructions

The main sources of obstruction have been described and remedies suggested. The following obstructions are not collided with but are in danger of becoming tangled within the machine. The main problems are the temporary services.

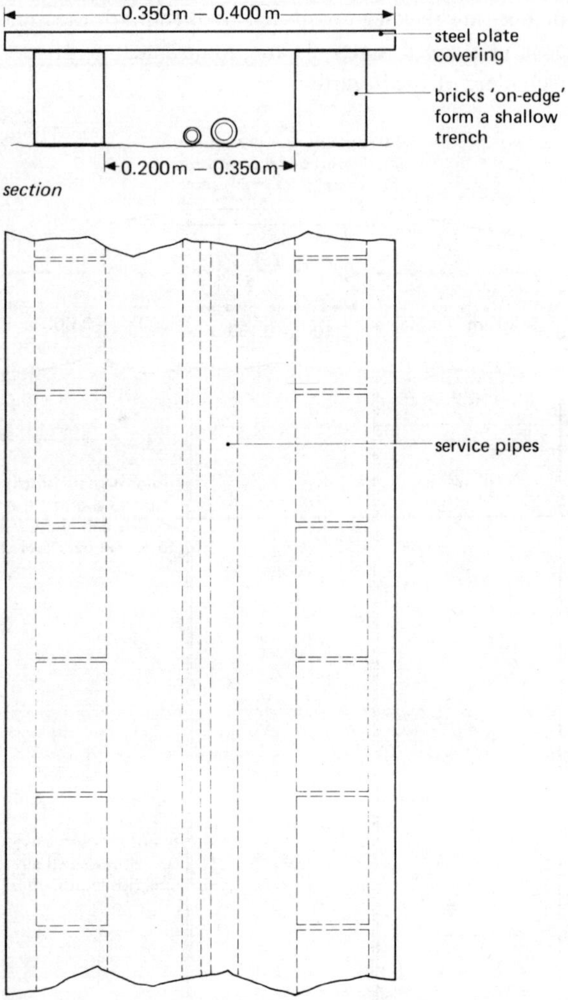

Figure 77 *Temporary duct for site services distribution. A shallow excavation is lined with sand on its base and bricks along its sides. The services are laid in this small, lined trench, which is covered with a chequered plate strong enough to carry site traffic*

Services at ground level

Where temporary supplies of electricity, water or compressed air pass across the site roadway, they must be protected. Burying them in a trench with sheet metal covering forms a suitable duct. If this is not practicable, then temporary ramps over the services must be provided. Under no circumstances must site transport travel over unprotected service pipes or cables.

Services above ground level

Temporary services above ground, e.g. telephone and electricity, must be clearly marked, to show their location and minimum headroom. Temporary 'goal-posts' are best but bunting flags ensure all site transport is safe. The principle is to erect a temporary, semi-rigid obstruction *below* the real obstruction, so that it is the temporary obstruction which is hit: the real obstruction is not threatened (Figure 78).

Machine error and failure

Definition

The machine is dependent upon the skill of its designer, manufacturer, operator and mechanic. It is not the machine itself that is in error, but the failure of human personnel to allow it to function correctly. The Construction (General Provisions) Regulations 1961 insist that all vehicles must be in efficient working order and good repair. This is achieved by the following:

Planned maintenance

The record of machine activities is scheduled so that services and maintenance can be organized when required, after so many working hours, or so many working weeks. By planning the work carried out on a machine, breakdowns should not occur. The breakdown or failure may be in a safety device, and so should be especially guarded against. A schedule of work carried out must be kept to ensure that future work is co-ordinated to maintain good safety standards.

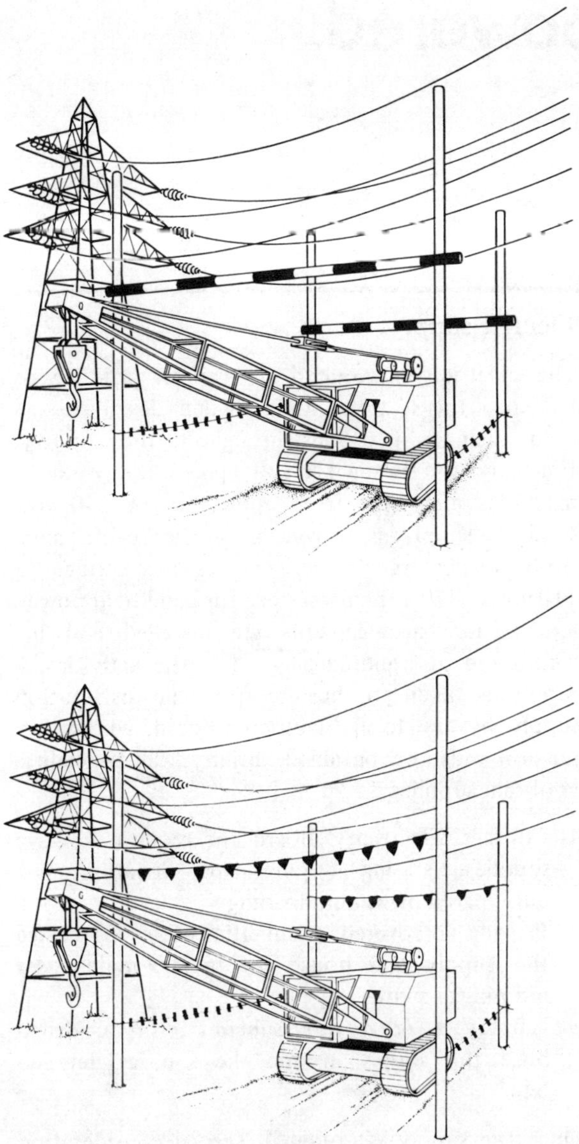

Figure 78 *Prevention of dangers from overhead power lines*

Preventive maintenance

If completed accurately, an operative maintenance scheme eliminates the need for planned maintenance activities. Whenever the machine is expected to be on a reduced workload a mechanical check is arranged. If this cannot be anticipated, the repairs and service maintenance must be completed outside normal working hours to maintain safety requirements.

Regular checking

It is important that the driver or operator maintains a regular check on his machine. A daily check on fuel, oil, water, tyres, guards, and all safety devices should be kept. A weekly check, by the driver or a trained fitter, should again look for any defect, especially those regarding safety.

Miscellaneous dangers

Fuel supplies

The location and uses of fuel storage are dealt with on pages 160-6. Restriction on quantities of fuel and prohibition of naked flame applies to all fuel storage. All those who fill transport vehicles from the fuel store area must comply.

Public highway

Where site transport is used on the highway a current Road Fund Tax disc must be displayed. In addition all requirements of the Road Traffic Act and licences for drivers must be complied with.

Unattended vehicles

All hydraulic or similar-powered equipment must be lowered to ground level when not in use. Engines must be turned off if left unattended.

Hired equipment

Certain conditions of the contractual agreement present special difficulties. The hirer (user) is responsible for safe working and the subsequent mechanical condition of the machine on a daily basis. The owner does not usually allow the hirer's personnel to adjust or maintain his equipment. The problem thus arising must be agreed between both parties and contractually bound to retain all necessary safe working of all site operatives.

14 Electricity and powered hand–tools

Electricity is an invaluable aid to site productivity. It is at the same time the most serious hazard to all construction workers because it is invisible and instantaneous in its effect. This chapter aims to advise the reader of how a site's temporary electric supply is organized and how it should be used.

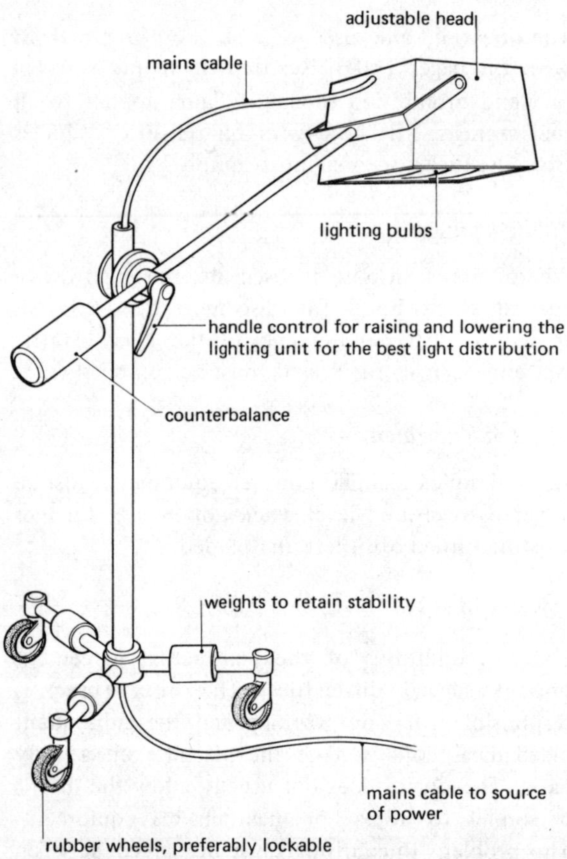

Figure 79 *Temporary site light*

Electric supply

The electricity generated by various processes is fed into the National Grid, which distributes an even balance of power throughout the country. For workshop uses, a 415 volt 3 phase supply powers machines and large fixed appliances. A 240 volt 1 phase supply is relayed to smaller plant, and, from this, a transformer reduces the voltage to 110 volts, 1 or 3 phase supply for hand-held power tools. These arrangements are installed and incorporated at building stage. For site activities, a supply is taken to the site from the distribution supply by the local Electricity Board, whose permission must be obtained. Briefly, the procedure to obtain supplies is:

1 Contact Electricity Board for request to have supplies, stating requirements, location and anticipated maximum loading
2 Provide a covered termination box to receive the supply and house the meter, main fuses and master switch
3 Employ a *bona fide* subcontractor to distribute the supply around the site, as detailed later

Once the supply is obtained, special consideration must be given to maintaining it safely and constantly throughout the site. The following factors are important:

Supply about site

The main contractor's incoming unit should be established at the edge of site activities, well clear of any risk of damage from plant or equipment. From this point distribution cables, at the correct voltage (Figure 33, page 83), cover the site.

R & D Building Company

Pararad Road, Roselip

STORES ISSUE OF POWERED TOOLS

Date	Appliance reference	Appliance	Plug	Cable	Conns.	Equipment received in good condition

Figure 80 *Form for issue of powered tools from store*

Supplies underground

All cables must be at least 0.500 m below the ground surface, with care taken to maintain this if uneven or undulating ground exists. All cables laid for a temporary supply must be protected within the trench by sand and be carefully covered. Every cableway must be marked with clear markers, and site operatives advised of these.

Supplies overhead

Uninsulated cables, which are usually used, create a very serious hazard to all metal objects, e.g. crane jibs, hydraulic diggers, etc. To avoid this, cables must not pass over roadways or areas where such transport may pass below. Where this is not practical or economic, temporary barriers in the form of goal-posts (Figure 78) must be erected. These provide a safe line to which all transport must be lowered to avoid the higher positioned hazard. Conspicuous markers of bunting flags or fixed rails are best for this work.

Existing supplies

These should be protected both above and below ground level as described in the preceding paragraphs and also on page 87.

Reduced voltage

The following chart indicates the requirements for reduced voltage on the site:

Heavy plant –	fixed, e.g. static crane, batching plant	415 volt	3 phase
Heavy plant –	unfixed, e.g. mobile tower crane, monorail requiring trailing cables	415 volt	3 phase
Hutting,	e.g. canteen, office, drying room, stores, toilets	240 volt	1 phase
Lightweight plant – fixed, e.g. circular saw, barbender		240 volt	1 phase
Fixed site floodlights, e.g. beam lights to large areas from high position		240 volt	1 phase
Lightweight plant – unfixed, e.g. concrete mixer		110 volt	1 phase
Unfixed site lighting, e.g. beam lights on movable tower at lower level		110 volt	1 phase
Portable power tools, e.g. drills, circular saws, sanders		110 volt	1 phase
Portable hand-lamps, e.g. general site use		110 volt	1 phase
Portable hand-lamps, e.g. confined areas or moist conditions		50 or 25 volt	1 phase

The reduced voltage is achieved by double-wound transformers which control the output. A range of sizes is available, each having outlets compatible with the transformer size.

Cables

All wiring must comply with IEE *Wiring Regulations*. The procedure to protect wires has already been considered, but where cables are required to lie on the ground, a safe ramp or barrier must be used for protection. The following cables are available for site conditions:

PVC (polyvinyl chloride) – suitable for permanent types of *internal* installation which are fixed, e.g. in canteen, hutting, stores

TRS (tough rubber sheath) – unsuitable for contact with solvents or oils but otherwise a strong type that withstands abrasive and similar abuse

PCP (polychlorapren) – resistant to abrasion, solvents and acids but retaining flexibility and suitable for all general purposes

Plugs

A standard form of coupling supplies is achieved by complying with BS 4343. Make-shift connections, not conforming, must not be attempted. Plugs and sockets must be connected properly and maintained regularly by a competent electrician. Where different phase wires are to be used the sockets are designed with different pin positions for single or three phase systems. These must not be connected incorrectly, and a colour code is incorporated as an additional safeguard.

Inspection and maintenance

The main contractor on site is responsible for the electrical procedures, site electrical plant or apparatus used. The installation, including wiring and connections to all equipment, must be inspected weekly. Accurate records, in writing, must be kept showing compliance with this requirement. Failure to complete these inspections makes all users and others on site vulnerable to serious injury in the event of malfunction. In addition, there must be a good maintenance procedure with records kept of all work and checks completed.

Earthing

Where a fault occurs the current travels back to earth by an 'earth loop'. Usually the impedance of the leak is low enough to activate the fuse. If this is not the case a serious risk of shock or fire exists. Reduction voltage transformers usually control the power, but these must also be correctly earthed.

Issue of appliances

In addition to providing a good electrical installation on site, the main contractor should ensure that all equipment from stores is safe. Every time an appliance is issued it should be visibly checked for defect, and the receiver should sign acceptance only after this inspection (Figure 80). This ensures that both employer and employee accept their responsibilities prior to using an electrical appliance.

Electric drill

This is one of the most common power tools, with a variety of uses within all trades. The most usual style has a pistol grip with a centre grip chuck incorporating a selection of speeds. The following considerations are for the safe operation of electric drills:

Double insulation

This gives double protection from electric shock by providing a complete external cover to all parts of the appliance. The international symbol shown above confirms compliance with this.

Switch lock

A locking pin within the handle retains the appliance in the ON position. This must never be secured and only used with extreme caution.

Percussion drill

Several types of drill have a percussion attachment built into their construction, which subjects the bit to regular light blows, during rotation. It is used for hard masonry drilling.

Speed selection

Using the correct speed on any drill gives a better job and, more important, is safer for the operator. Charts of suitable speed settings are shown in Figure 81, page 138.

Good practices

1. Always tighten the chuck securely
2. The workpiece must always be secured, unless large enough to avoid movement
3. Maintain carbon brushes regularly. Excessive sparking indicates this deficiency or a possible short circuit
4. Take advantage of the speed selections. Start drilling into all hard materials at a slow speed and increase accordingly
5. Slow down the rate of feed just prior to breaking through the surface, particularly of metals, to avoid snatching or twisting
6. When drilling large masonry holes, use a pilot cutter for initial drilling
7. Keep drill vents clear to maintain adequate ventilation
8. Use sharp drills at all times
9. Make sure all plug connections and fuses are secure and correct
10. Keep all cables clear of the cutting area during use

Bad practices

1. Never use a three-wire drill on two wires only: *always connect on earth*
2. Do not use a drill bit with a bent spindle
3. Do not exceed the manufacturer's recommended maximum capacities for drilling sizes on appropriate material
4. Never use a hole saw cutter without the pilot cutter
5. Do not use High Speed Steel (HSS) bits without cooling or lubrication

Two-speed units

1st speed – selector knob – Position 1
2nd speed – selector knob – Position 2

Material	*Drill diameter*	*Speed*
Wood	2 – 28mm	2
	28 – 35mm	1
Steel	2 – 11mm	2
	11 – 13mm	1
Aluminium	2 – 15mm	2
	15 – 17mm	1
Masonry	2 – 10mm	2
	10 – 28mm	1

Four-speed units

1st speed – selector knob – Position 1
switch trigger – Position 1
2nd speed – selector knob – Position 1
switch trigger – Position 2
3rd speed – selector knob – Position 2
switch trigger – Position 1
4th speed – selector knob – Position 2
switch trigger – Position 2

Material	*Drill diameter*	*Speed*
Wood	2 – 16mm	4
	16 – 28mm	3
	28 – 32mm	2
	32 – 35mm	1
Steel	2 – 8mm	4
	8 – 11mm	3
	11 – 12mm	2
	12 – 13mm	1
Aluminium	2 – 10mm	4
	10 – 15mm	3
	15 – 16mm	2
	16 – 17mm	1
Masonry	2 – 6mm	4
	6 – 10mm	3
	10 – 20mm	2
	20 – 28mm	1

Figure 81 *Recommended speeds for electric drills – two-speed and four-speed*

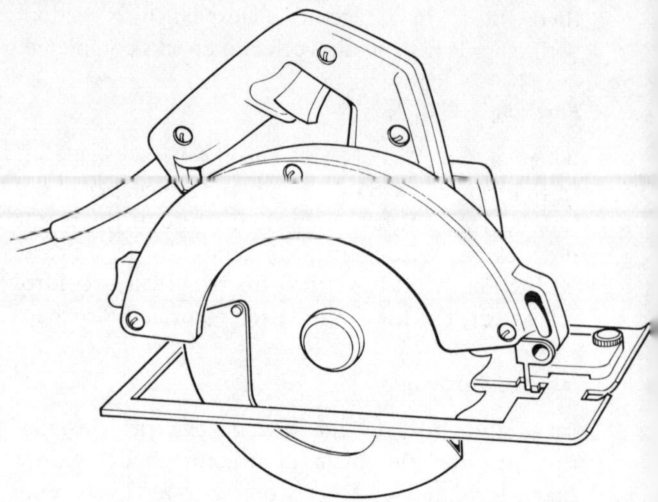

Figure 82 *Hand-held powered circular saw*

Hand-controlled circular saws

Used mainly by carpenters, the saws combine the quick and easy cutting of timber with the advantages of being easy to manoeuvre and light to handle. The following are considerations of safe use and activities with circular saws of various sizes and types:

Extension cables

The manufacturer's recommendations, of three core cable, 13 amp of tough rubber sheath, should be observed and nothing inferior be used.

Operation

Always use the correct blade for the work involved, e.g. rip or crosscut, and ensure the teeth are in the direction of the arrows on the saw guards. Retain a firm grip without force. Always handle the lower guard with the knob provided.

Overload slipping clutch

Most saws have a device that allows the motor to continue rotating if the blade has stalled under pressure. If this pressure cannot be released quickly, release the switch trigger.

Good practices

1 Wear eyeshields as necessary, particularly when cutting aluminium
2 Wear eyeshields, protective clothing and breathing apparatus when cutting asbestos
3 Always use a sharp blade that is appropriate for the work to be done
4 Allow the machine to attain full revs before cutting the material
5 Regularly blow out dust from the motor and check carbon brushes for wear or replacement
6 Always make sure the lower guard is fully returned before laying down the saw. Most saws have automatic retractible guards

Bad practices

1 Never allow a hand to be placed under the sole plate or guard of the machine
2 Do not adjust or change the blade whilst the saw is connected to the mains
3 Do not overtighten the locking nut
4 Never allow cable to be in front of the cutting line
5 Never carry or drag the saw by the electric cable
6 Do not hold or fix the guard in the open position
7 Do not twist the saw in an attempt to check alignment
8 Never use a saw that vibrates or appears unsatisfactory in any way
9 Do not force the saw at any time during cutting
10 Do not cut materials without first checking for obstructions or foreign objects, e.g. nails, screws

Electric sabre saws (jig saws)

A very versatile joiner's powered tool used for cutting shapes in relatively thin materials. They are used on site for *in situ* cutting. The following factors should be considered whilst using one of these saws:

Operation

Allow the saw adequate time to cut completely without force or haste. Avoid forcing the machine to keep alignment and do not overtax the cutter in an attempt to achieve deep cutting.

Good practices

1 Always allow the machine to turn with ease, do not force it around a curve
2 Always change or adjust blades with the electric supply disconnected
3 Select the correct blade and allow it to cut steadily
4 Stop and allow the blade to become stationary before withdrawing it from the workpiece
5 Use lubricants when cutting metals, oil for cutting mild steel, paraffin for cutting aluminium

Bad practices

1 Do not force the tool along or bend it around the curved cutting line
2 Never allow the cable to be in front of the cutter during use
3 Do not commence cutting until the machine reaches its full revs
4 Always position the machine before cutting, and avoid re-entry with an activated blade

Powered belt sanders

These are most frequently used by carpenters, joiners and painters to clean off surfaces, usually

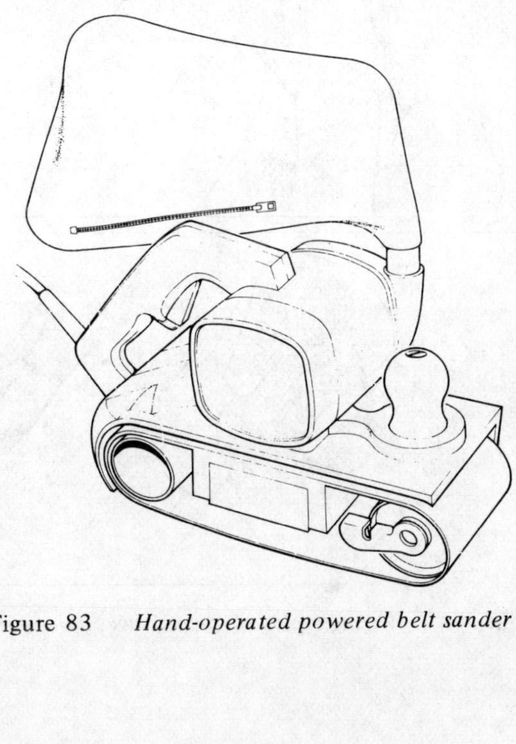

Figure 83 *Hand-operated powered belt sander*

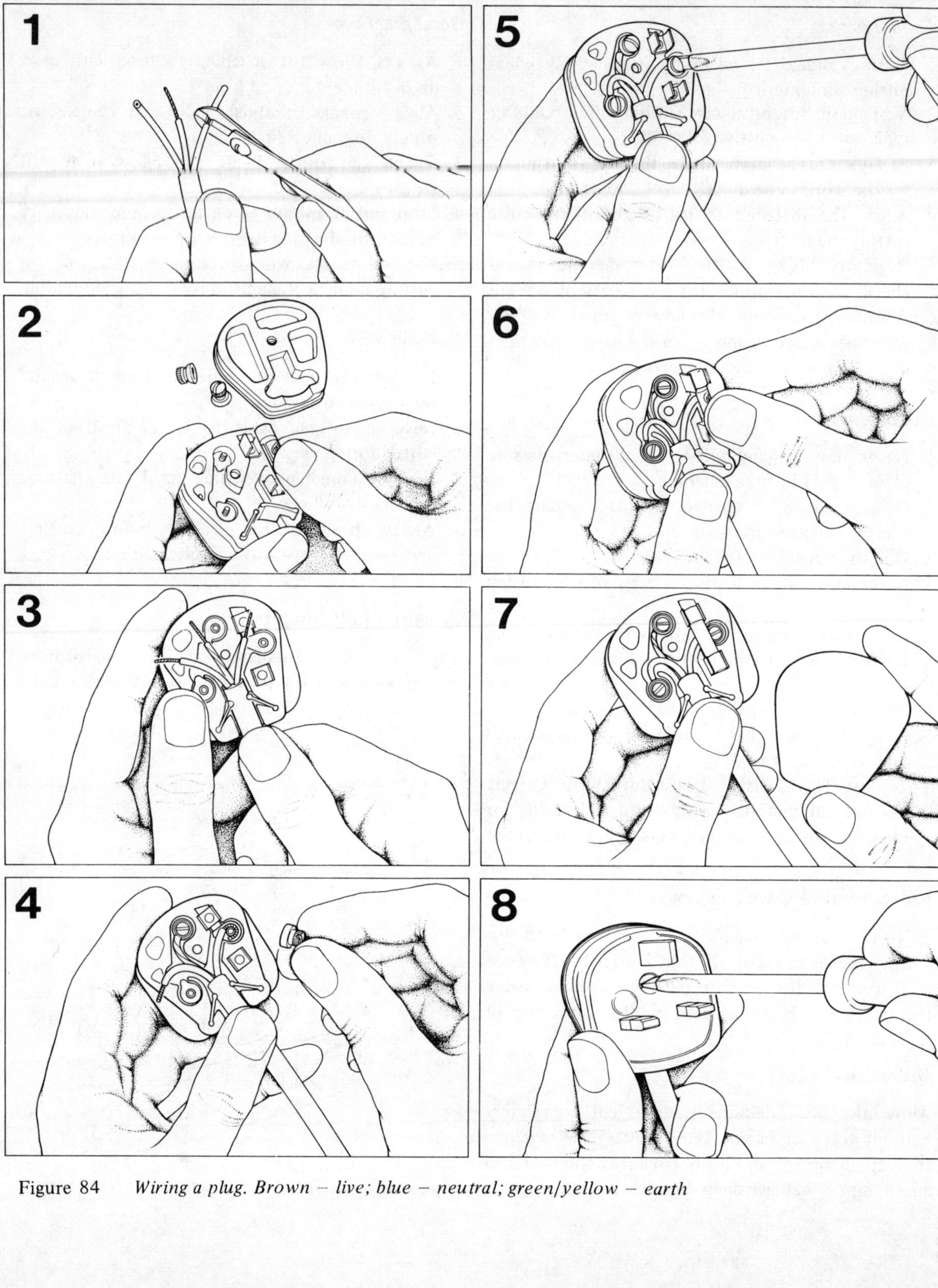

Figure 84 *Wiring a plug. Brown – live; blue – neutral; green/yellow – earth*

timber, to receive decoration. Regardless of the trade or capacity of the user these rules should be followed:

Operation

The machine should be allowed to run at full speed, in a forward position and with a firm grip. Additional weight is not necessary and may cause damage to the machine or workpiece. Certain types of sander will not operate unless the vacuum bag is attached. Failure to secure this bag allows dust particles to be discharged at high velocity, which is dangerous.

Good practices

1 Secure the abrasive belt in the correct direction as indicated on the belt and machine
2 Clean dust from motor at regular intervals
3 Keep the vacuum bag attached at all times and empty the contents when it is a quarter filled
4 Keep cables well clear of working surface
5 Secure the workpiece or use a 'stop block' to prevent movement or ejection of small work-pieces which may cause injury to others

Bad practices

1 Never attempt to change abrasive belts, or empty the vacuum bag whilst the sander is connected to mains supply
2 Do not exert excessive pressure upon the moving machine
3 Do not work on unfixed materials, unless they are of sufficient weight to remain static. Other persons are easily injured by flying parts ejected by sanders
4 Never cover the air vents
5 Never allow the cable to be trapped under the belt. Keep the cable clear of the working surface

Plugs and fuses

All plugs must be correctly wired to terminals and all cables must be held firmly by the cord grip and the plug cover engaged. This applies to *all* powered hand-tools mentioned (Figure 84).

Fuses must always be used; improvised fusing by silver paper or nails is most dangerous and must never be attempted.

15 Cartridge tools

The use of cartridge-powered tools has increased in recent years. This type of tool created a new method of fixture that dispensed with electrical or compressed air power, with their cables, hoses and similar restrictions. Being portable, the cartridge-powered tool can work in confined areas, be independent and very versatile.

The principle of cartridge-powered tools is that an explosive cartridge is loaded into a barrel behind a fixing device. The explosion from the cartridge activates a piston which drives the fixing device down the barrel and into position. Tests already completed confirm that serious personal injury can occur up to 450 m – more than a quarter of a mile – away. This makes the appliance potentially lethal and worthy of extreme respect.

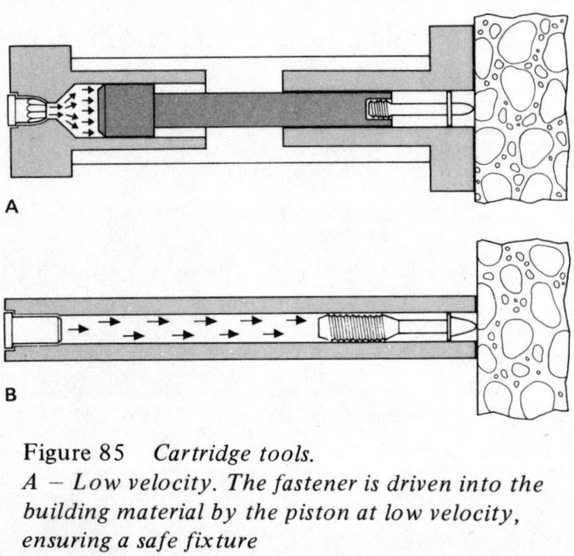

Figure 85 *Cartridge tools.*
A – Low velocity. The fastener is driven into the building material by the piston at low velocity, ensuring a safe fixture
B – High velocity. The fastener is driven along the barrel and impacts with the building material at high velocity

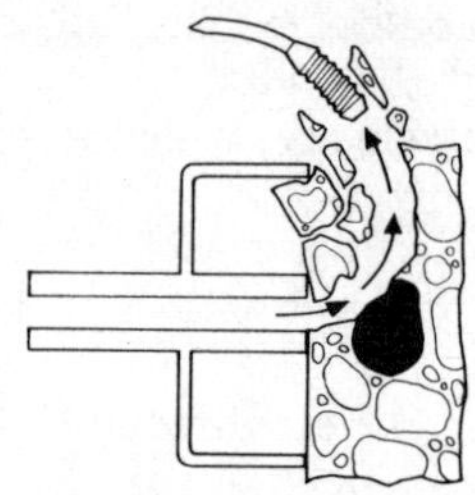

Ricochets or fishhooking may occur when driving fasteners on concrete. Reinforcing rods or aggregate increase this hazard.

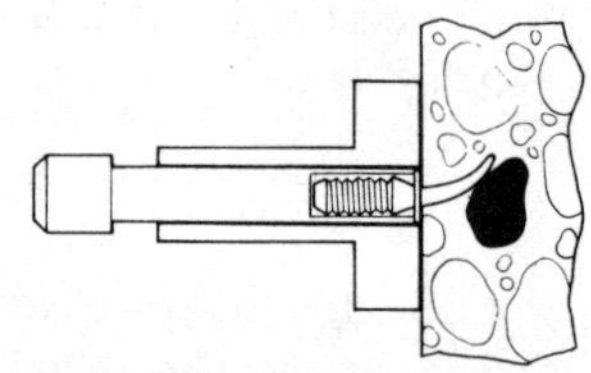

Since the driving force on the fastener ceases immediately when th piston is stopped inside the tool, th fastener is deflected but stays inside the material.

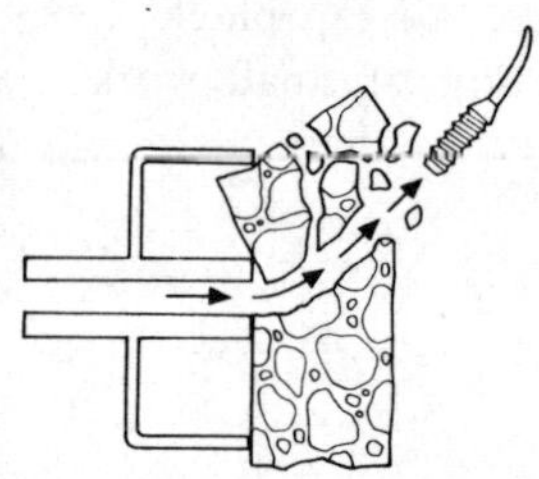

If an edge breaks away or spalls while driving a fastener into concrete, the fastener may be dangerously deflected to one side.

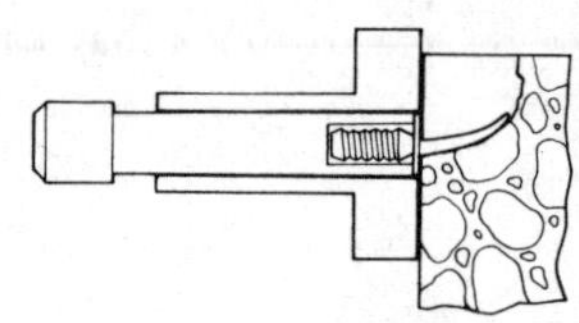

Even if an edge does spall there is nc ricochet hazard because the piston i stopped and the fastener possesses only a minimum of residual driving energy.

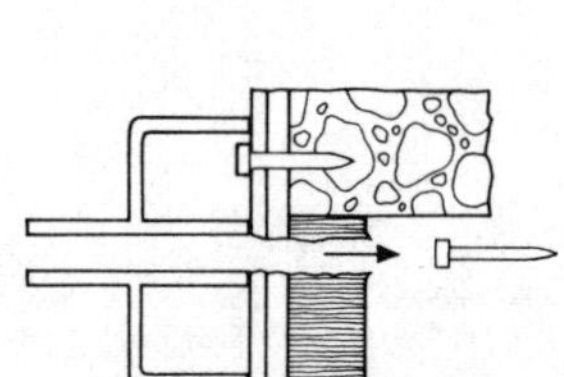

If the base material is not tested or there is a hidden change from concrete to lightweight construction a fastener could be shot through a wall in free flight.

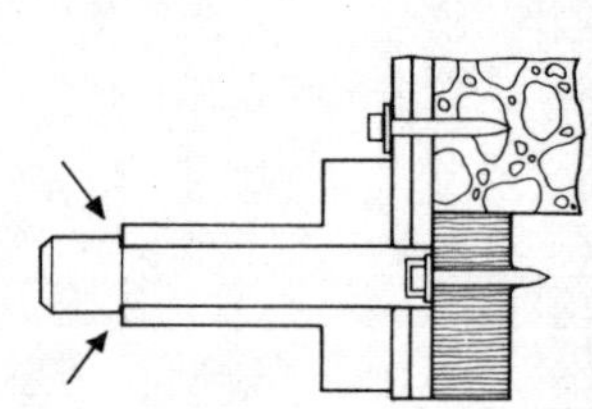

Piston controlled penetration alm fully eliminates dangerous throug shots: when the piston stops so dc the fastener.

Figure 86 *Advantages of the low-velocity piston cartridge tool*

R & D Building Company

Pararad Road, Roselip

Site Address/Workshop

Issue & Distribution of Dangerous Equipment

Date	Equipment Issued	Equipment reference No.	Signature of receiver	Equipment Returned	Signature of returner

Figure 87 *Form for issue of dangerous equipment*

The cartridges are coloured according to their force and power in accordance with the following code (see also page 144):

Extra low – brown
Low – green
Low/medium – yellow
Medium – blue
Medium/high – red
High – white
Extra high – black

'Fixing device' is a general term which covers nails, thread-headed fixtures, etc., that are supplied in various forms by the manufacturer.

The storage of cartridge tools and their cartridges needs caution and precision to avoid any misunderstanding or misuse. A good distribution scheme must eliminate the possibility of careless issue or a failure to return unused equipment. A typical example of a stores distribution procedure is shown in Figure 87. All equipment must be returned to the store at the end of use or of the working day, whichever is sooner, to avoid it falling into incompetent or dangerous hands. It is important that the guns and cartridges are stored separately in a cool, dry, locked cabinet with a security arrangement that reduces careless or indiscriminate issue.

2.10 *Cartridges*

2.10.1 Cartridges designed for use in a cartridge-operated tool shall be marked clearly with the strength of the charge.

2.10.2 Only cartridges corresponding to specifications issued by the maker of the tool shall be used.

2.10.3 The bottom of the cartridge shall show the manufacturer's mark. The strength of the cartridge shall be distinguished by a colour on each cartridge. The colour identification code shall be as follows:

Cartridge strength	Code letter (s)	Identification colour	Colour No. in BS 381C*
Extra low	XL	Brown	410
Low	L	Green	217
Low/Medium	LM	Yellow	309
Medium	M	Blue	166
Medium/High	MH	Red	537
High	H	White	–
Extra high	XH	Black	–

*BS 381C, 'Colours for specific purposes'.

2.10.4 The cartridges shall be packed in containers with coloured labels, the colour corresponding to the cartridge contents, and the container shall show clearly the following information.

1 The tool manufacturer's name or trade mark.
2 The words 'Safety cartridge'.
3 Calibre.
4 Strength of cartridge by colour and code letters.
5 The words 'only for . . . tool'.

Figure 88 *Extract from BS 4078 concerning cartridge tools*

Preparation and personal protection

The preparation for use is based mainly upon protection of the body against injury whilst using the appliance.

Head

Where splintering is expected, head injuries may be caused by deflected, brittle materials, a safety helmet must be worn.

Eyes

In accordance with the Protection of Eyes Regulations, goggles to BS 2092 must be worn.

Ears

The explosion causes excessive noise levels at a very close range which may damage the user's ears. Ear protection must be worn whenever cartridge-powered tools are used.

Flammable

Cartridge tools must not be used where a flammable atmosphere exists. The cartridge explosion may cause fire or further explosion of ignited gases.

Guarding

Before use, the apparatus must be checked for a suitable guarding device. Most types of cartridge tool are designed to be inoperative if the face guard is removed; the equipment will only function if the guard is placed flush to the working surface. The surface area of the guard covers the impact area to reduce splintering.

Work activity, hazards and precautions

The user of cartridge tools must be at least eighteen years old, unless under personal supervision or training. They should have completed a training schedule and been issued with a certificate of competence. This training is easily available from most manufacturers and offers a good preparation against the following possible hazards.

Any operative who may use a cartridge tool must have good eyesight and must not suffer from colour blindness. There are responsibilities for both employer and employee to eliminate this error by careful testing before initial employment.

Misfire

When the appliance malfunctions, the operative should refire the trigger without withdrawing from the firing position. Should a second failure occur the operator must retain the firing position for at least fifteen seconds before removal. At this stage, great care is needed to point the barrel in a harmless direction and take out the offending cartridge in strict accordance with the manufacturer's instructions. All defective cartridges must be stored separately and returned to the manufacturer for inspection, testing and destruction.

Recoil

The thrust inflicted upon the operator by the appliance may cause unbalancing. To resist this a firm stance is required. It is important to control the strength of cartridge used in particularly hazardous situations where recoil may be dangerous, e.g. on ladders.

Action

To activate the gun, its protective guard must be pressed firmly against the workpiece. It is important to ensure that the structure is not easily penetrated and that the guard does not allow firing into open spaces.

Ricochet

In certain cases the fixing device can bend under force as it enters the structural fabric. This can be caused by a hard, dense particle, or by a hole created by a previous, unsuccessful fixture. This effectively redirects the fixing device back towards the operator. The danger is greater where high-powered tools are used or if materials shatter outwards. This problem can be avoided or reduced by using only low-powered tools, and making all attempts to fix at least 0.050 m away from previous fixtures or future attempts.

Splintering

Where brittle fabrics need to be penetrated, splintering will occur if any resistance is met with against fixture. The operator's body must be protected as previously described, and he should have a clear, shielded area around him with warning notices erected.

Maintenance

Whenever malfunction or misfire becomes frequent, the appliance must be returned to its manufacturer for checking and correction. It is good practice to have all cartridge tools overhauled regularly, at least annually.

16 Compressed air

Compressed air generates power for pneumatic tools. There is much to be said in favour of powered tools, which reduce the labour of the operatives. There is also a great deal to be said regarding safe working practices with compressed air.

The compressor

The motor that provides the compressed air needs regular maintenance in the same way as any other plant. In addition to normal mechanical and electrical failure, there are other hazards. The compressor must always be positioned in a well-ventilated area clear of any obstruction that may reduce a clear passage of air to it. A firm base should always be found on which to stand the compressor unit and a notice erected warning site personnel to keep clear. The machine should only be activated or controlled by operatives well acquainted with its actions. Training should be given to all operatives, and no one under the age of eighteen should use the compressor or pneumatic tools. A responsible person must take charge of the compressor and be competent to inspect all parts of the equipment including plant, hoses, couplings and tools. This competent person must have the qualities described on page 35. He should inspect the equipment regularly and include a report of repairs required and any maintenance that has been completed.

The connections and hoses

The size of hose is compatible with the pneumatic tool. Where the bore of the pipe varies there will be a variation of power; the compressor should be positioned close to the work so the hose length is short and there is no snaking and tripping. If a long hose is necessary, a large-bore hose should be used, with a 'shut-off' valve fitted near to the tool, which reduces the hose to a bore compatible with that of the required tool. The hose must be kept clean, especially when disconnected or stored. All leaks must be repaired, and any form of corrosive material or chemical must be kept clear of hoses and couplings. The filter within the hose line must be well cleaned and drained to avoid a drop in pressure. It is important to prevent any vehicle from crossing over the hose, particularly whilst it is in use. It is equally dangerous and foolish to bend the hose to reduce power. Either of these actions creates unnecessary disturbance to the flow of compressed air. By squeezing the hose, extra pressure is exerted on to the couplings which may fail. When a coupling or connection fails the hose whips, coils and turns at great speed, causing a severe blow to anything in its path and seriously injuring any part of the body. Regular maintenance by a competent person is important. All defects must be remedied and weakened parts of the hose or couplings replaced immediately.

Protective garments

Pages 69 and 96-101 indicate the degree of protection required. Generally air tools produce more noise than electrically powered, and ear protection is important. Pneumatic tools create dangers to the eyes, making the use of goggles necessary. The use of gloves and protective footwear is advisable but varies according to the job.

General safety procedures

When the air line is free from the appliance, it produces jets of air which may be foolishly misused. It cannot be over-emphasized that compressed air, if allowed to enter the bloodstream, causes painful swelling that can be fatal. Similarly, serious injury may result where compressed air hits the ears, eyes or nose. The following checklist should be observed for all compressed-air tools.

1 Never indulge in horseplay or do anything foolish with compressed air or its apparatus
2 Never direct compressed air at anyone
3 Do not use compressed air to clean away metal swarf, wood chippings or any material that might create flying particles
4 Never dust your clothing or that of anyone else with the charged airline
5 Do not use compressed air to create a draught to prime or aid heaters and stoves where a risk of fire may occur
6 Never use a compressed-air tool until the maker's instructions are fully understood
7 Never use a compressed-air tool without correct pressures
8 Ensure a regularly cleaned filter is fitted within the hoseline to avoid a drop in pressure
9 Always have a silencer or muffler on all appliances
10 Allow all operators to practice with the tool before prolonged use

Rotary tools

Power for rotary tools is provided by the air motor which drives a spindle directly or through a gearing system. Either an electric or vane motor can be used; the air motor has the advantage of being reversible. There are no dangers of burn-out or fire if the appliance is stalled, overloaded or strained.

Abrasive tools

The abrasive appliance is carried upon a shaft extension of an air motor. The regulations insist that abrasive wheels exeeding 0.055 m diameter (including disc sanding tools) must be clearly marked by the manufacturer with the maximum speed for the wheel. A marking must also show the maximum speed for any tool, and air-driven equipment must be governed to keep within the maximum permissible speed. It is important that all abrasive wheel tools are guarded, preferably with an adjustable all-purpose guard. Those abrasive tools that require extraction of particles (grinders, sanders, etc.) should be fitted with a dust extractor to reduce health hazards. Some form of vacuum or suction device should be used to draw the unwanted dust into a dust-tight bag.

Drills

All drills work to the maximum diameter bits specified by the manufacturer. The user must always respect the size and capability of the drill and never tax it beyond its designed limits. It is important to use the best grip to control the equipment. A choice of hold is available, a short 'pistol' grip for smaller types and a two-handed grip, sometimes with a chest plate, for larger drills. The user must always ensure a good secure hold upon the drill cutter, remembering to use the correct chuck key to tighten the hold.

Saws

A maximum saw blade of 0.225 m is permissible and must only be fitted by a competent person over the age of eighteen. Adjustable guards are fitted and must always be used as the manufacturer intended. Wedging back of the guard for repetitive work must not be permitted. The user must be trained in its fundamental safety needs, including fluent running, observing regulations and maintaining the tool in good order.

Screw-drivers and impact wrenches

As they are similar to the drill, these rotary devices require the same safety considerations. An additional requirement is that, on applying the screw-driver, special attention is needed to avoid unnecessary labour of the motor. The slower, geared-down equipment works by forward pressure upon it; if the operator does not produce this pressure the machine may malfunction and subsequently create a hazard.

Wood-boring equipment

Being a version of the drill, this equipment again requires the same safety practices. Special care is needed when using the reversing gear for withdrawing or holding off from the work. It is important to choose the correct gear for the work to be done, e.g. a smaller cutter requires a slower gearing to retain compatibility between motor and cutter.

Percussive tools

Compressed air generates the backwards and forwards movements of a piston, at high speed, within a cylindrical frame. The hammer-like blows are exerted on to the internal end of the cutting or breaking tool which activates the lower, bladed end.

Breakers

A wide range of tools varying in shape, robustness, consumption of compressed air and degree of strength. The safety factors however do not vary. All breakers must be fitted with silencer and/or mufflers to control the excessive noise. All operators must wear ear protectors whilst using these tools, and in many cases goggles should also be worn. It is important that all operators wear the correct footwear: this cannot be over-emphasized (pages 99-100). Strong steel-toe-capped boots only guard the front part of the foot. The operator should have good quality boots to protect as much as possible of his foot and maintain extra care and concentration. Suitable gloves should be worn to minimize cuts, blisters and bruises. The vibrations set up by the pneumatic breaker cannot be eliminated, but the operator must not be subjected to constant use of vibrating tools. Substantial rest periods must be taken to avoid physical or muscular disorders.

Hammer tools

A smaller version of the preceding tools, used for drilling, chasing and cutting. Some versions include a rotary action. The same risks apply as for the breakers and identical safety activities must be applied.

17 Posture, fatigue and body damage

The science of body movements is referred to as 'kinetics'. Posture is the position, at a given moment, of the various parts of the body in co-ordination for performing a set task. This does not imply an erect or straight body but one which allows the correct position to undertake the work.

Fatigue and several subsidiary ailments are caused through postural strain and lack of consideration for the body condition. The modern trend towards mechanization has reduced the physical stress exerted upon the body but has highlighted problem areas. Technology has often made work more repetitive, resulting in the stagnation of many body muscles. When the movements are restricted to a few repetitive actions, the efficiency of blood vessels is reduced and body fatigue increases. Tension in certain body tissues retards absorption by compressing blood vessels.

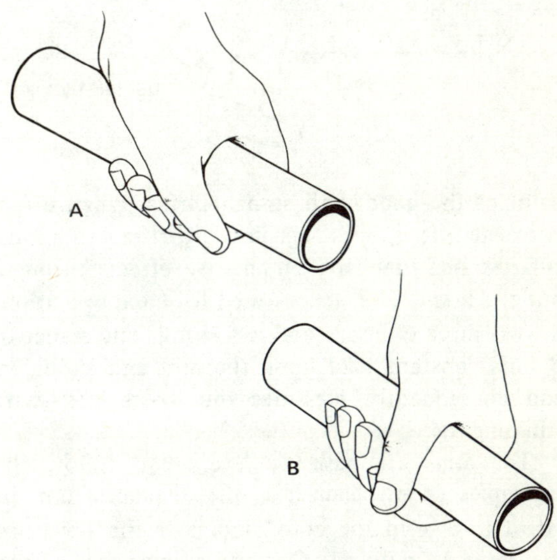

Figure 89 *Hand gripping action*

Arms and hands

The close bond and functional relationship between the arm, forearm, wrist and hand creates an intricate system with thousands of uses. The effort needed by the body is influenced by the holding and lifting actions of the hand and arm. If pressure is put upon the ends of the fingers, contraction of the bending muscles throughout the body causes tension. If, however, effort is exerted on to the palm of the hand, less strain is imposed upon the forearms, arm and the chest, so there is less fatigue. A good example is the grip used on tools, e.g. shovel, pick, scaffold tube. A typical gripping action is shown in Figure 89. On the left, A shows extra effort being induced because pressure to the finger ends creates tension to arms, shoulders and back. The alternative in B shows the base of the fingers being used to advantage, and less body effort is created. It is good working practice, and an additional safety aid, to grip tools from the underside where the object is lifted by the palm of the hand. From this position the possibility of dropping the tool, or object is greatly reduced.

The insecurity of hold created by fingertip grip causes a change of carrying position and may induce unnecessary and exhausting strain upon muscles. The error of many people is to hold the object high, which is more strenuous and less secure.

The correct position can be achieved (Figure 91) with a firm grip at the base of the fingers or

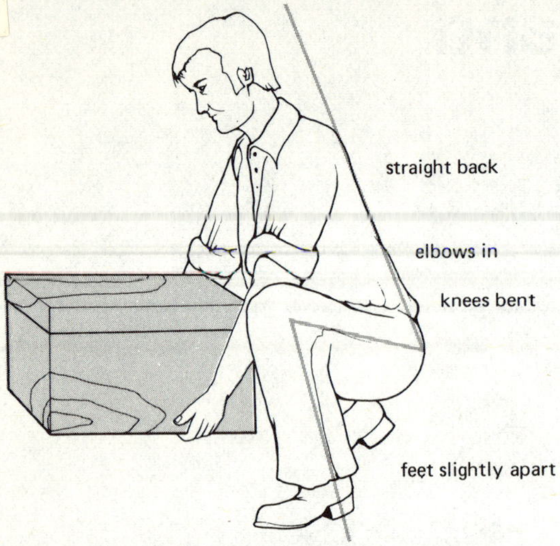

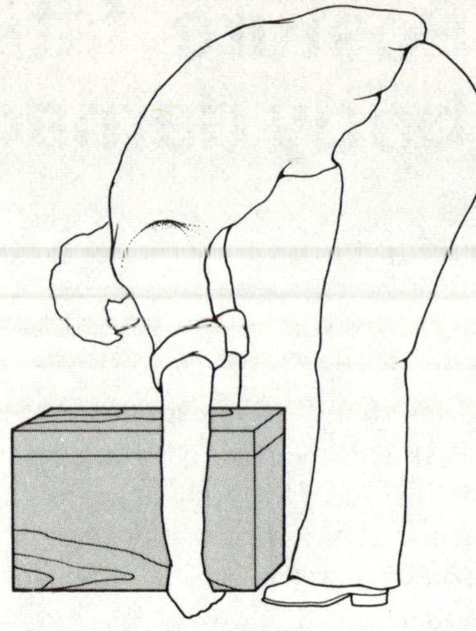

Figure 90 *Correct and incorrect lifting*
A – Correct. Straight back and bent legs
B – Incorrect. Bent back and straight legs

palm of the hand with straight arms. When effort is exerted, it is often needless effort creating muscular strains that have cumulative effects throughout the body. The stances used for hand operations in A is more effective and less tiring. The stance of B puts constant load upon the arm and shoulders and consequently pressures the wrists into extra straining force.

The arms are basically levers and follow the principles of mechanics; as the distance from the effort (i.e. from the body) increases the power of the lever is reduced. Operatives must be trained to avoid unnecessary bodily pressures and the 'brute force' approach which so often causes damage.

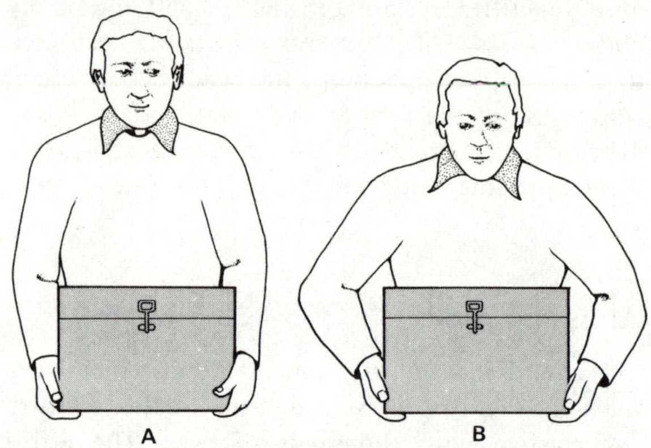

Figure 91 *Stances used for hand operations.*
A – More effective and less tiring
B – Puts constant load on arms and shoulders

Back

The back is most vulnerable to bad lifting practices. It is at its best lifting position when kept straight, the force coming from the thigh and leg muscles which are designed for lifting. The back is most exposed to damage when weakened by bending forward. Figure 90 shows good lifting technique with the back kept straight. The approach is to stand with one foot slightly forward to enable the backward foot to push forward in the movement direction. The inter-relationship of arms, back, and legs co-ordinates the lift, projects the body forward and causes the minimum of body damage. By using the lower, stronger limbs the person controls the effort and maintains a good body condition for future work activities. Figure 90 also shows an incorrect lifting procedure where the back is bent and weakened. A great pressure is put upon the lower limbs and back muscles and the body is unbalanced on legs kept parallel. The upper muscles of these lower limbs are therefore pressured, causing danger and fatigue to the body of the lifter.

Chin

A minor but vital part of the lifting procedure is to position the chin correctly. By keeping the chin in towards the chest, the spine is retained in its more natural upright shape. Subsequent pressure upon the back is reduced and the leg muscles used more effectively.

Legs and feet

The legs act as adjustable props which regulate the stability of the body. Additional support is given by the feet, even to the fine degree of the toes' gripping actions within the operative's footwear. A lack of consideration for the legs results in back injuries created by cumulative strain. There is a danger of the body being unbalanced and muscles of the lower limbs contracting to compensate. Foot position is important and allowance for the bodily 'follow through' must be considered to avoid strain upon the back. A high proportion of industrial ailments and strains are caused by a lack of knowledge of good body positions. Training should include advice upon this topic at an early stage of the employee's career.

Another important function of the lower limbs is to aid the body from collapse should an obstruction occur. If a foot stumbles or is tripped, nerves flash information through the body to form a spontaneous response to stop a collapse. If a foot feels a slight tilt, the nervous system reacts to control the body. The speed of this reaction influences body movements and avoids falls. A sluggish response does not stop the body from falling, whereas a quicker body reaction keeps the victim upright.

Conclusion

The most serious reason for injuries caused by lifting is personal ignorance of good posture. The dominance on site of those who can show the greatest power is a serious problem. Many individuals attempt loads in excess of their physical capabilities and often lift them incorrectly. Employees must be discouraged from this unnecessary activity and be trained to take advantage of mechanical lifting appliances. Where such equipment may be uneconomical or impossible to install the employee must still avoid straining his body. Incorrect working and lifting techniques create fatigue which reduce the work rate and probably create a hazardous situation. The following checklist applies in each situation where physical damage may occur:

1. Always size up and assess the job before lifting, and make sure it is within your capabilities
2. Ensure a good grip and correct posture
3. Make sure the back is straight and not overburdened
4. Always keep a clear vision whilst lifting
5. Use lifting hooks and aids where applicable
6. Do not over-reach to stack materials
7. Prepare and check the route to avoid any slippery surface or obstacles
8. Stop and re-lift, in preference to maintaining a lift, if delay occurs
9. Always wear protective clothing, e.g. boots with steel toe caps, in case grip is lost and the load dropped
10. Take precautions to avoid damage from dangerous materials, e.g. acids and chemicals

18 Hand-tools

Hand-tools are those that involve no source of power – electricity, compressed air, etc. – apart from the user's own strength. Although hand-tools do not immediately seem as dangerous as power tools, they must be used properly.

Hitting tools

Hammers are the basic tool for hitting. Some hammers also act as levers to withdraw nails. In making them, the poll or striking end is heat-treated and quenched in water. The lever end of a claw hammer is heat-treated and quenched in oil. This process reduces metal fatigue and provides a ductile, tough tool. Where hammers are used only for striking, heat treatment is followed by water quenching. Handles may be of wood, steel, or fibreglass. The vital part of hammer construction is the joint between the handle shaft and the head. This is the most vulnerable failure point. Wood handles should be made from ash or hickory incorporating a straight grain of mature growth and moisture content between 10 and 14 per cent. The joint is best achieved with a single wooden wedge and two, toothed, malleable iron wedges. Metal handles are made from hardened and tempered steel with rubber grip handles. The joint between shaft and head is made permanent by an 'interference' fit achieved by hydraulic driving. The fibreglass shaft is drawn through the head and permanently bonded with epoxy resin.

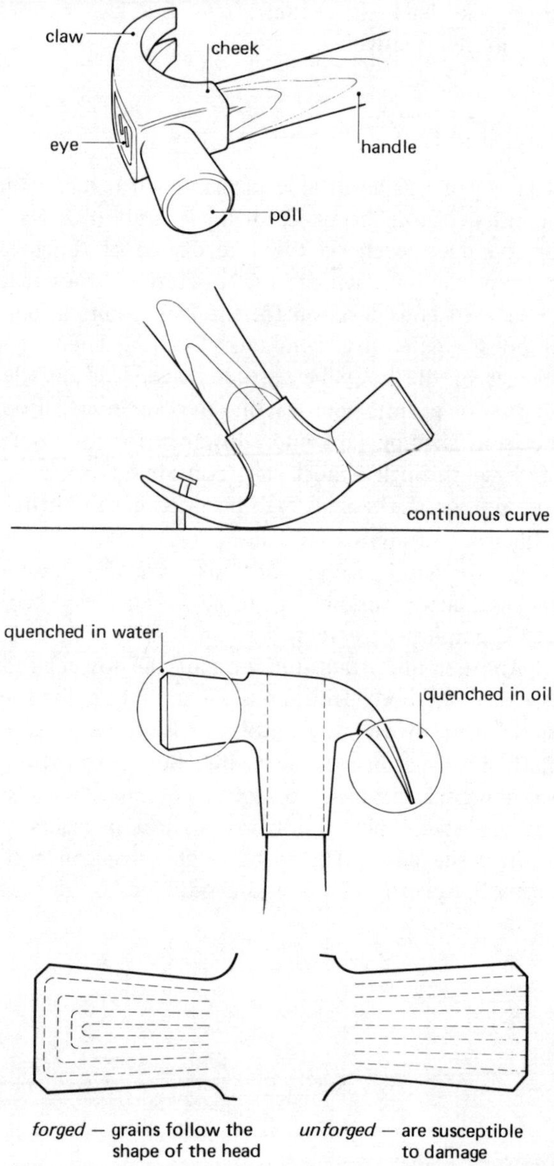

Figure 92 *Hammer manufacture and uses. An unforged hammer head has an open grain structure, and is easily damaged if hard objects or other hammer heads are struck*

Hammers include the following:

Claw hammer, for hitting and withdrawing nails
Warrington (cross pein), for lighter duties. The pein is for starting small or inaccessible nails
Engineers (ball pein), for striking at the poll end and riveting at the ball pein
Pin hammers, a small version of the Warrington for light duties
Brick hammers, for cutting bricks accurately and chipping bricks and brickwork to detailed requirements
Club hammers, heavy-duty hammers for cutting bricks or masonry in conjunction with masonry chisels
Soft hammers/mallets, especially prepared in lead, copper, plastic or rubber, required for the dressing of malleable metals, etc.

Regardless of the manufacture, type or use of any hammer, safety demands the following:

1 Strike squarely and avoid glancing blows
2 Do not strike with the side of any hammer
3 Never strike two hammer faces together (tempered steel shatters dangerously)
4 Always wear eye protection where the work is hazardous

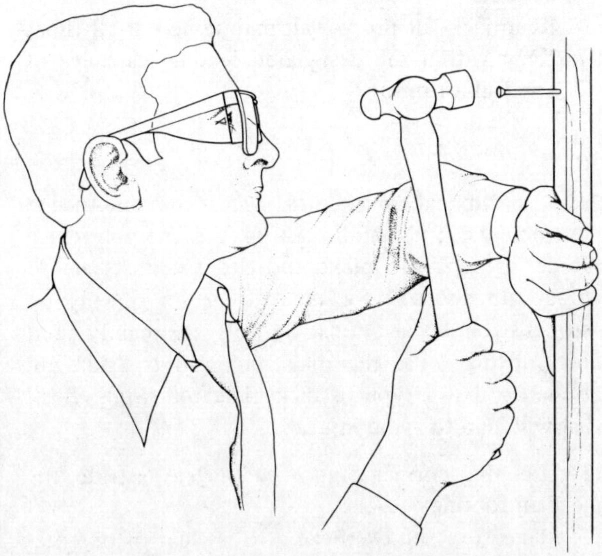

Figure 93 *Fixing masonry nails. The operator wears suitable eye protectors and holds the workpiece well away from the area near the masonry nail*

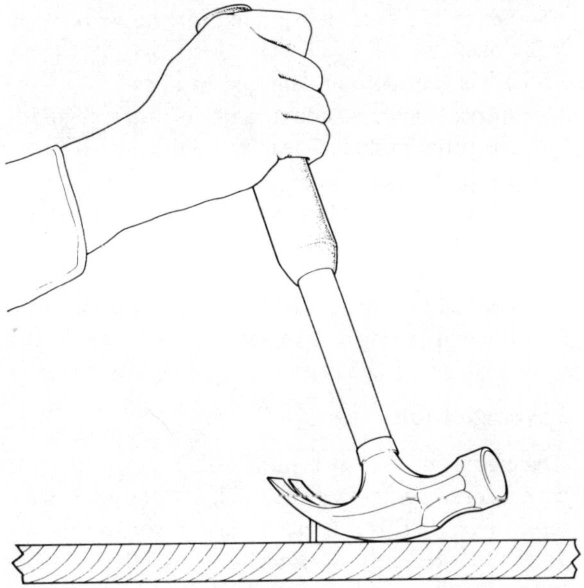

Figure 94 *Using a claw hammer to remove nails*

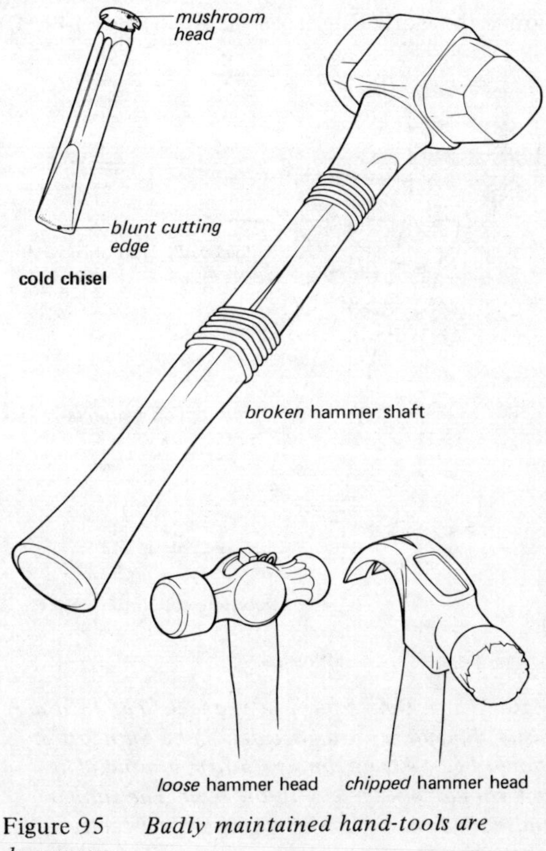

Figure 95 *Badly maintained hand-tools are dangerous*

5 Select the correct hammer for the work to be done
6 Never use loose or damaged handles
7 Inspect and maintain any hammer regularly
8 Keep the poll face clean by rubbing with emery cloth or sandpaper
9 Never strike metals harder than those of the hammer head or the hammer head may shatter dangerously
10 Use the hammer sensibly – do not use excessive force in hitting or leverage

Leverage tools

Leverage is movement in terms of force – upwards with jacks, across with a pick, or around with screwdrivers. This latter, rotary force is known as torque. Hazards in these tools are the grip between the tool and the material, and the joint between head and handle shaft. A further consideration is the worker's grip on the tool (pages 149-50).

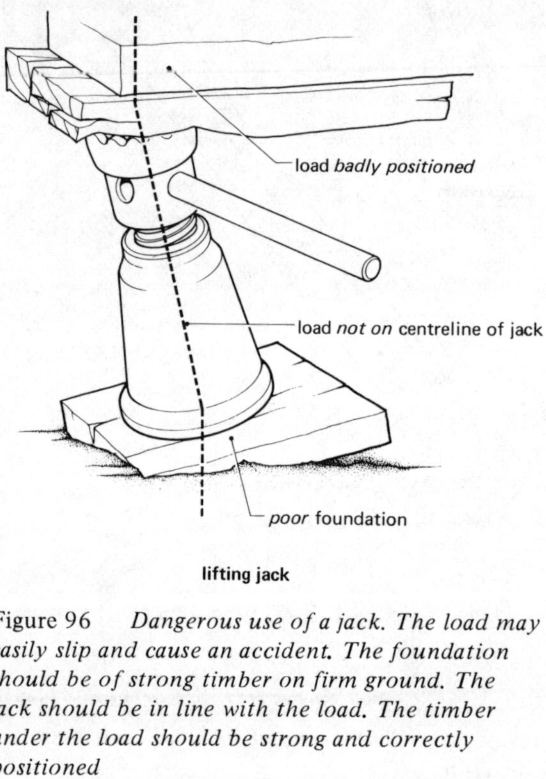

Figure 96 *Dangerous use of a jack. The load may easily slip and cause an accident. The foundation should be of strong timber on firm ground. The jack should be in line with the load. The timber under the load should be strong and correctly positioned*

Levers forcing up

The following are the main tools or appliances for this type of work:

Hydraulic jack – for lifting temporary loads
Adjustable steel props – for temporary load-bearing use (adjustable trench props, although used horizontally, are much the same)

The first safety factor is that support must be strong enough. Short cuts and improvising must be avoided. The following rules should always be observed:

1 Always support the load centrally on the jack, to retain concentric loading and avoid eccentric loading (Figure 96)
2 In the case of jacks, always 'block up' alongside to avoid collapse
3 In the case of props, always secure the head and foot plate to timber supports (sole plate and head)
4 Never support from bumpy or unstable ground
5 Never improvise. This is particularly important with adjustable steel props when it is dangerous to use mild steel bars or nails as pins. The manufactured pins of high-yield steel must always be used
6 Keep jacks or props well maintained at all times
7 Keep within the designated loading capacity of any jack or prop

Levers forcing across

The actions taken by these levers are side-to-side movements, the handle giving the main leverage force. Typical are picks and shovels. Picks are to break up the soil for removal or for digging up stones from excavations. Shovels are mainly used for moving loose materials and also to 'side up' trenches like a type of pick. The following check list will help to avoid injuries:

1 Use the correct stance, with legs astride for firm footing
2 Make sure the overhead area is clear of obstructions
3 Keep picks sharp with suitable heat treatment to retain a hardened point

4 Always keep picks and shovels well maintained, their handles free from splinters, cracks or similar defects, and their metal parts clean and free from splits
5 Do not leave any tools lying on the ground. They might be a tripping hazard or become damaged

Levers forcing horizontally

Where the force pulls horizontally the tool usually acts as a lever against a fixed item, e.g. spanners, wrenches and pliers. The greatest area of error is the grip between the tool and the object:

1 Always use the correct size of jaws to avoid slipping and subsequent hand injuries (Figure 97)
2 Never attempt to increase the leverage by adding sleeved additions to the spanner length. This will strip off threads or snap bolts and studs (Figure 98)
3 Do not hit any spanner or wrench with a hammer
4 Never try to use two spanners with open ends interlocked
5 Do not interchange tools, e.g. never use pliers instead of spanners

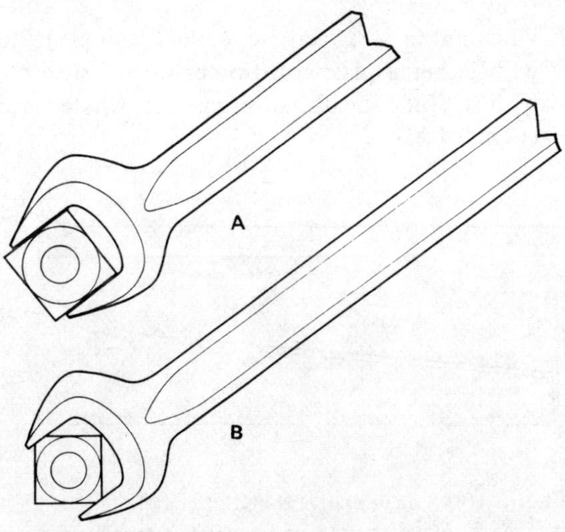

Figure 97 *Use the right spanner.*
A – Right. The spanner fits tightly and cannot slip
B – Wrong. The spanner is a loose fit and can slip off. Even if it does not, it damages the nut by burring the edges

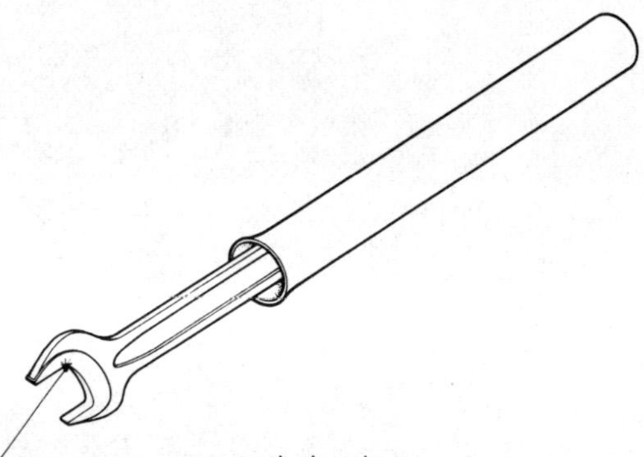

Figure 98 *How a spanner is strained. It is tempting to use metal tubing on a spanner to increase the torque and put extra pressure on the jaws. Weakening and fracture of the spanner's jaws may occur, as well as damage to the nut and bolt threads*

6 Take extra care when cutting tensioned wire as the end tends to fly off
7 Pliers for electrical use must have insulated handles
8 All leverage tools must be well maintained and kept to the correct adjustment

Torque

Another style of leverage tool is the screwdriver, where the user's effort is exerted by a torque. Screwdrivers are available in a range of size and type of shank to fit different screw heads. Follow these safety suggestions:

1 Always use the correct size of shank for the screw head
2 Keep screwdrivers properly maintained with an accurately ground tip
3 Never carry screwdrivers in pockets as uncomfortable scratches or skin punctures will occur
4 Do not hammer screwdrivers
5 Never use a screwdriver for leverage or as a chisel to remove obstructions
6 Do not exert excessive pressures upon the work, especially where overbalancing may occur
7 Screwdrivers for electrical work must be insulated
8 'Yankee' screwdrivers must never be ejected carelessly. Always open these out slowly

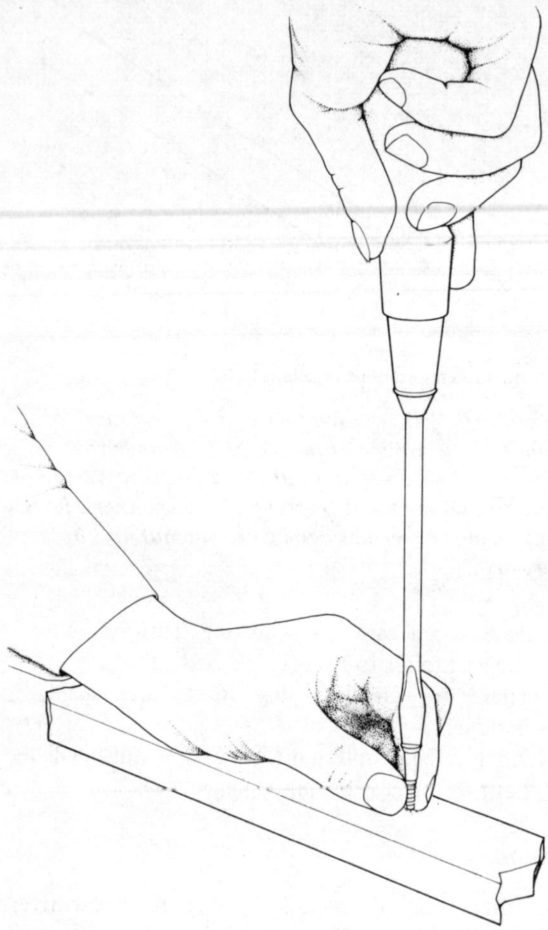

Figure 99 *Correct use of a screwdriver*

9 Never point the yankee screwdriver towards anyone prior to ejecting the spiral
10 Keep yankee types well away from children

Cutting tools

This range of tools is the largest and most various; at the same time it is the most dangerous. Regardless of its use, any cutting tool needs a sharp cutting edge. Any edge that can cut, sever or break materials can easily damage human skin and tissues.

Timber-cutting tools

This range of tools, mainly used by carpenters and joiners, must have sharp cutting edges, to cut accurately all timbers, which in the main are comparatively soft, workable materials. Blunt edges require additional force from the user, perhaps resulting in injury.

The following rules apply to the use of chisels:

1 Keep chisels sharpened to the correct angle, grinding angle 20-25 degrees, sharpening angle 30-35 degrees
2 Use the right type of chisel for the job
3 Where chips may be hazardous, wear eye protection
4 Always keep hands and body behind the cutting edge
5 Store chisels with a 'storage roll', a cloth bag with slots for each chisel which folds and rolls for complete protection
6 During the working activities, when not actually using the chisel, place it safely or replace the plastic protective cap on the cutting edge
7 Always use a mallet for all chisels; never hit them with a hammer

When using a saw observe these rules:

1 Use the correct type of saw for the job
2 Make sure the teeth are properly set and adjusted to avoid binding
3 When not in use, keep the teeth of saws protected with either a plastic or timber slotted sheath or with a full sheath covering the whole blade (Figure 100)

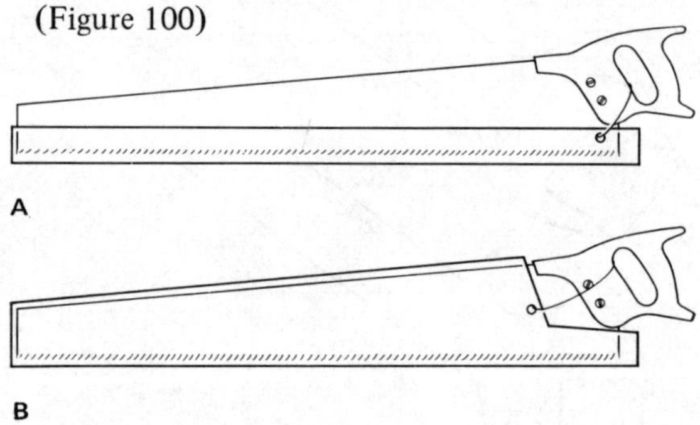

Figure 100 *Saw protectors.*
A – Part sleeve protector covers the teeth, preventing cuts from the sharp edges. It is held in place by a short tie through the handle
B – Complete sleeve gives better protection. This type generally holds itself in position, but a short tie can also be used

Figure 101 *Use vice and tenon saw correctly.*
A – Right. The workpiece is held securely in the vice near the part to be worked on. Hands are not in vulnerable places
B – Wrong. The workpiece is held in the vice too far from the part to be worked on, and is not secure. The operator holds on to the workpiece in an attempt to steady it and is easily cut if the saw slips

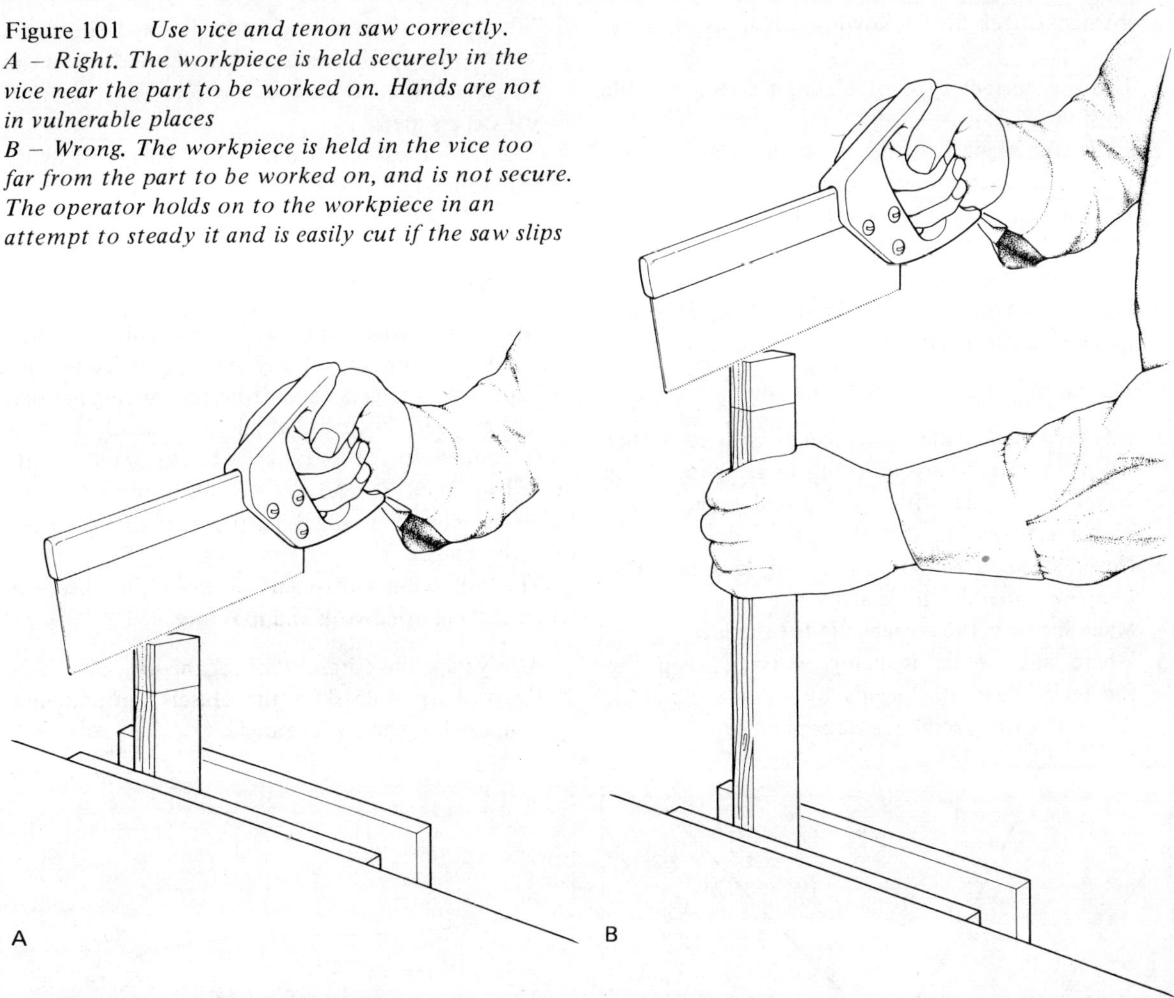

A B

4 Always protect the saw from rust after contact with any moisture, by smearing with oil
5 Avoid excessive, forceful use where the saw may buckle and cause injury
6 Where binding may occur, use wedges to avoid any dangers
7 Do not work where offcuts might fall on others without giving warning

Use the following checklist for the safe use of knives:

1 Use the correct type of knife
2 Make sure the work is on a flat surface and the workpiece secure
3 When not in use, all knives must be stored within a sheath. An ideal solution to this is use only knives with retractable blades
4 Where several knives are used continually in a workshop, a protective rack should be installed
5 Interchangeable blades, small blades and razor blades must only be used in properly designed holders
6 Keep all edges sharp, excess pressure will often cause accidents

Metal-cutting tools

Many different types of metal are available, all with varying qualities and characteristics. Generally a softer metal can be cut with snips or hacksaws, which create no real safety hazard. Harder, or more dense metals offer more resistance and may create problems. Check the following when using hack-saws:

1 Use the correct type of blade, set to a suitable tension
2 Insert the blade with the teeth pointing forward
3 Apply sufficient power to cut steadily with full strokes along the length of the blade
4 Unless the workpiece is large and self-stabilizing, a suitable fixture must be used
5 Do not over-use a worn blade; it is safer and quicker to replace it

Check the following for metalwork files:

1 Always make the workpiece secure, either within a vice or by blocking larger workpieces
2 Never use a file with an exposed tang; always have a handle secured to the file
3 Never strike a file or use it for leverage. The tempered metal will shatter
4 Make sure it is the correct file for the job
5 Where soft metal is being worked, keep the file teeth clear of clogging by using a file cord. A clogged file becomes smooth and slips
6 Where hard metals are being worked the teeth wear quickly becoming smooth and slippery. To avoid this, select the correct file, or grind off excess metal
7 Keep all files free from oil or grease to avoid slipperiness

Brickwork and masonry-cutting tools

Although the bricklayer uses a wide range of hand tools, only a few create any hazard whilst in use. The principal dangers come from the cutting actions of chisels and bolsters which are struck with a club hammer to cut away, split bricks or chase out, etc. The main dangers come from mushroomed heads which allow small particles of metal to fly off dangerously.

The following safe practices should be observed when cutting brickwork and masonry:

1 Always use the correct tool for the job
2 Keep sharp points on the chisels, ground and tempered to the correct angle

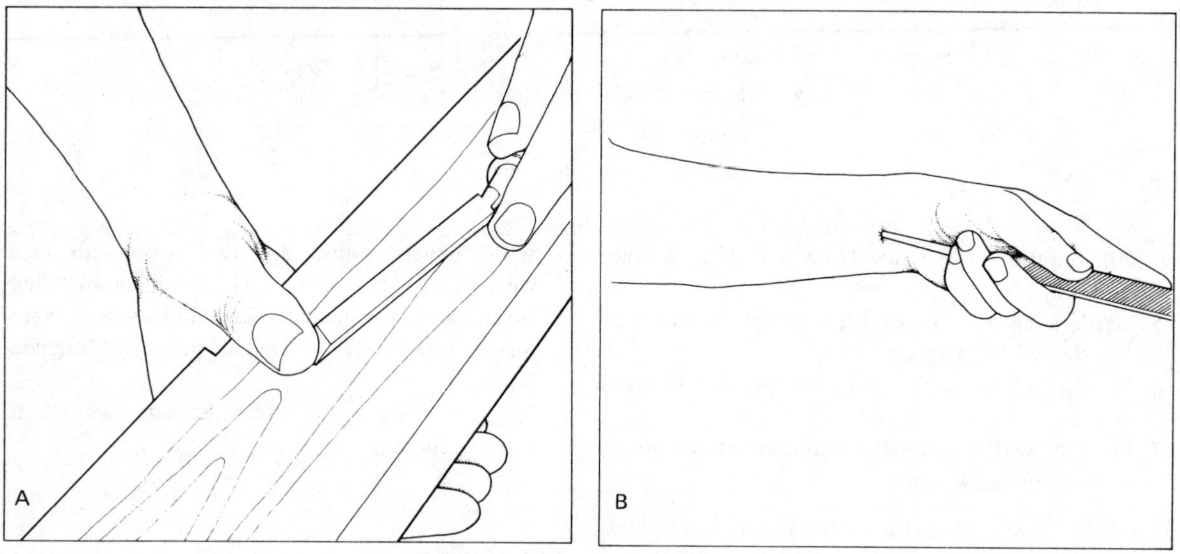

Figure 102 *Wood chisels cut . . . and so can files.*
A – You must keep both hands behind a cutting tool. It should never be necessary to work with one hand in front. Blunt cutting tools may be a contributory factor as they require the operator to exert excessive pressure
B – This file was used without the correct handle to prevent the tang cutting into the operator's hand. There has been a failure by the employer to provide suitable tools and by the employee to take reasonable care

3 Never allow the equipment to have a mushroomed head. If this does occur, regrind to a slight taper (Figure 95, page 153)
4 Where necessary use eye protection. This is specified within the context of the Protection of Eyes Regulations (pages 98 and 169)
5 Work in a systematic way, striking away from the body to reduce injury from flying particles

Shaping tools

Shaping tools are those used by carpenters and joiners for planing and shaping timber or plastic. These act on a similar principle: a base through which a sharpened blade projects. The hazards usually occur when sharpening and replacing plane irons. The following is a checklist for using shaping tools:

1 Use the correct equipment for the job
2 Keep all blades sharp and at the correct angle. To use excess pressure might mean the user slips or over-reaches
3 Never hold plane irons where the screwdriver may puncture the hand beneath. It is better to support the fitment on a flat surface (Figure 103)
4 Always dismantle a plane over a workbench to avoid the possibility of the blade dropping on to legs or feet
5 Store the apparatus with blades retracted and protected from damage or contact with moisture

Conclusion

The following general points should be considered when purchasing, using, or storing tools:

1 Avoid the false economy of cheap tools. These are often less well made and more easily damaged, easily mushroomed, quickly blunted, handles coming loose, and so on
2 Take special care to use the correct apparatus for the job. Particular attention is needed to use insulated tools near electricity
3 Take good care of all tools, with careful oiling to moving parts. Keep all equipment clean of unwanted dirt and grease.

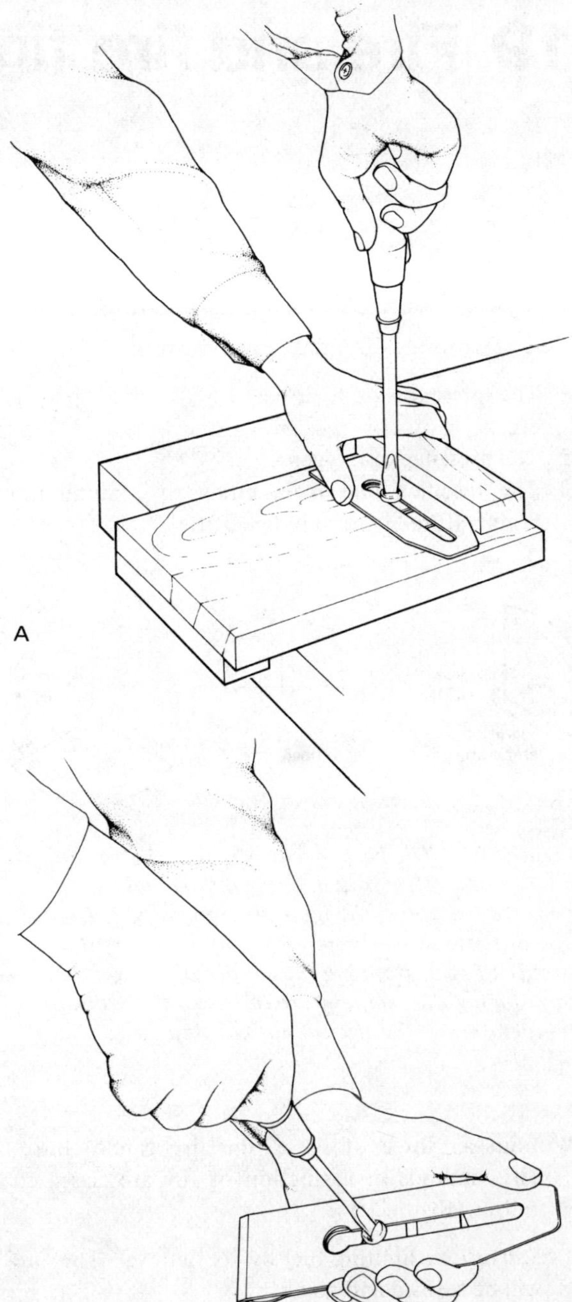

Figure 103 *Change a plane iron correctly.*
A – Right. The operator holds the plane firmly on the bench, and no damage to his hand is possible
B – Wrong. If the screwdriver slips, the operator's left hand is vulnerable

19 Fire and fire fighting

Three factors are essential to combustion:

1 The presence of fuel, e.g. wood, paper, petrol, etc.
2 The presence of oxygen
3 The reaching and maintaining of a minimum temperature 'ignition temperature'

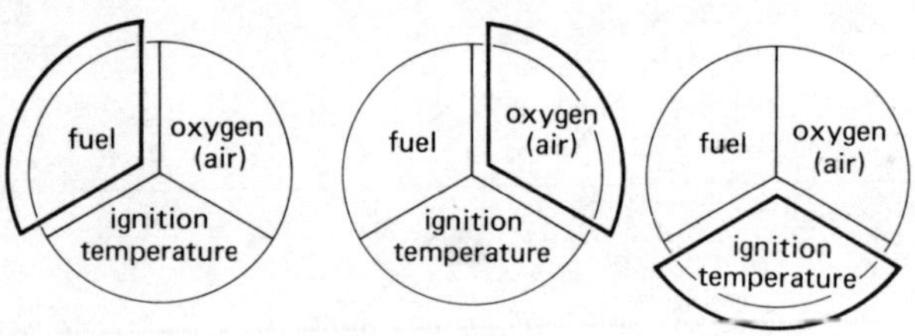

Figure 104 *The basic needs of a fire. A fire can only burn if there is fuel, oxygen (air) and a temperature above the ignition temperature. If any one of these three is removed – by stopping the supply of fuel, stopping the supply of oxygen (air), or reducing the temperature to below the ignition temperature – the fire cannot continue*

Without all three of these, the fire cannot burn.

The methods of extinction of fire are classified as follows (Figure 104):

1 *Starvation:* limiting fuel by its removal. The fire will be starved and go out
2 *Smothering:* the limitation of oxygen by foam extinguishers or asbestos blanket
3 *Cooling:* the limitation of temperature. Each fuel has a certain minimum ignition temperature

The application of water on carbonaceous fuels reduces the temperature of the fuel below its ignition temperature and extinguishes the fire. To be effective the water must be aimed at the *fuel* – the base of the fire and not the flames.

The prevention of fire in completed buildings is a vast and detailed topic which is beyond the realm of this book. This chapter highlights the hazards of site and workshop fires and suggests preventive actions.

Causes of fire

Fire is only possible where there is a supply of combustible materials in construction work or in a workshop. Lack of caution, equipment failure or simple human error can cause ignition.

Flammable liquids and their appliances

Storage: the nature of these liquids makes them very vulnerable to ignition and the following precautions as necessary:

1 All containers must be clearly marked, showing the contents
2 All containers must have secure capping devices
3 Flammable liquids must be stored in a securely locked compound at least 6.096 m from any other stores, hutting, site boundaries or structural buildings
4 Naked flames or lit cigarettes must be prohibited within 6.096 m of any liquid gas store
5 Where drums are used as containers, they must have taps with drip trays or drip cans
6 All empty containers must be returned to store or supplier as soon as possible
7 Transportation must only take place whilst containers are sealed

8 All empty containers must be stored on end
9 Spillages must be cleaned up or covered with sand

Appliances: any apparatus that requires a flammable liquid needs the basic precautionary measures as stored liquids. Any spillage must be cleaned up immediately or, if on the ground, covered with sand. All tank or machine leakages must be reported and then repaired immediately. Drivers' and operators must be warned of ignition hazards whilst refuelling; they must not smoke and the engines must be stopped.

Electrical appliances

Temporary electrical installations: a completely separate installation may be required by the contractor for the execution of his work. Even though it is temporary, it must be installed by a *bona fide* electrician. The following rules apply:

1 Isolation by a main switch and adequate protection against excess current must be included
2 All overhead wiring must be strong enough for the span and marker flags used for protection (page 133)
3 All ground wiring must be underground or protected from vehicular damage by the use of ramps (pages 131-2)
4 All portable electric-powered hand-tools must be connected with flexible rubber or sheathed PVC cables
5 Hand-held lamps with tungsten bulbs must be suitably protected from smashing without restricting the light
6 Power supplies must be reduced to a safe voltage (see also pages 82, 134-6)
7 Any machinery of mobile or static type must be fixed with isolating devices to cut off the machine during adjustments or alterations

Liquefied Petroleum Gas (LPG)

The dangers related to the storage, use, equipment and operation of these gases are reviewed on pages 201-11

Combustible materials

Rubbish is a severe problem in fire prevention, but it can always be avoided by good management. Any build-up of packing fabrics, wood offcuts and shavings, oil waste or off-cuts of combustible materials constitute a potential fire hazard. The following check list must always be observed:

1 All waste liquids that are flammable must be contained in sealed units for disposal off site, or burned in small quantities, in shallow trays
2 Site or workshop waste must be burned in a suitable incinerator. Where this is not practical, *small*, well-controlled fires can be used, providing these are at least 9.140 m from hutting, storage or any building structure
3 When open fires are ignited, they must be constantly supervised and damped down after combustion has ended and at the end of the day
4 Extinguishers must be available at, or near, any fire lit for the disposal of rubbish
5 All rags of an oily nature must be retained in metal bins with close-fitting sealed lids

Building structure and fabric cause similar problems in and around the building area, so consideration must be given to this checklist:

1 All containers of flammable liquids must be clearly marked and secured with tight-fitting capping devices
2 Any drums of flammable liquids must have a tap with a drip tray or drip can
3 Any spillage of flammable liquid must be cleaned up or covered with sand
4 All appliances and equipment using flammable liquids must be maintained in good condition, and all defects reported and repaired immediately
5 All wiring for electrical appliances must be protected with ramps from damage from foot traffic or moving plant
6 All electric tools and hand-held lamps must be connected with strong flexible cable and used with the correct voltage
7 All precautions related to welding operations and the gases used within the area are applicable (see pages 205-11)

8 Any combustible waste materials or liquids of a flammable nature need to be cleaned regularly and must not be allowed to accumulate within the building area
9 All burning of rubbish must be carried out under supervision at least 9.140 m away from the building area

Temporary hutting

Because hutting is only temporary, the contractor may be tempted to 'short cut' or eliminate certain safety obligations. There is a tendency to build hutting from combustible materials with elementary heating appliances and no regard for fire hazard or means of escape. The following should be observed:

1 All hutting should be of non-combustible materials. If for practical reasons, timber huts are erected, they must have suitable fireproof lining
2 Huts must be well spaced out to eliminate the spread of fire from hut to hut. Each unit should be well clear of the ground. To eliminate vermin, a barrier should be fixed to prevent rubbish accumulating around or underneath the huts
3 Storage racks should be well spaced with 'fire-gaps' around them to prevent the spread of fire
4 Where huts are erected within the building, they must be of non-combustible material,
5 Clear 'No Smoking' notices must be erected to stop smoking in hazardous areas, e.g. carpenters' workshops, covered storage for combustibles, etc.
6 Where smoking is permissible, adequate supplies of ashtrays and sand buckets must be available and cleaned out regularly
7 Any heating apparatus must be installed so that it cannot be knocked over or damaged and become a fire hazard
8 All heaters, boilers, etc., must be installed upon a concrete base, with an ash well of 0.076 m, an asbestos backguard, and protection where the flue passes through the hut structure, to avoid fire
9 All heaters must be well maintained
10 Gases and all other fuels must be stored on the outside of the hut, store, or room where the appliance is installed

Planning for fire prevention

Three main stages of the construction process require special attention when planning the prevention of fire. These are pre-contract design, pre-contract planning for fire prevention, and the work process (site or workshop).

Pre-contract design

There are both moral and legal obligations upon planners, designers and engineers to create a high standard of fire prevention, within the general standards of all construction. The specified use of certain combustible materials often creates a hazardous situation and consultation with the Fire Prevention Department is important. The following checklist covers the major considerations:

1 All materials must comply with relevant British Standards and, where applicable, to any fire test required by the same British Standard and/or code of practice
2 All superficial treatments, e.g. impregnation of timber, must not reduce protection from fire
3 Every part of any building must have a means of escape from fire
4 Adequate means to control fire must be included, e.g. fire and smoke stop doors, fireproof barriers to flues and ducts, etc.
5 Adequate and suitable fire extinguishers must be installed throughout the building on completion
6 Suitable hoses and similar fire fighting aids must be installed and connected to supplies of water as a part of the construction
7 A complete, infallible alarm system must be installed

Pre-contract fire prevention planning

The successful contractor must complete several planning procedures before work begins. The following pre-contract planning applies to fire prevention:

1 Arrange a regular fire (and rescue) inspection by the local fire and rescue service

R & D Building Company

Pararad Road, Roselip

CONTRACT:

CONTRACT ADDRESS:

	HAZARD	ACTIONS REQUIRED	ACTION BY	REMARKS
MAIN SITE OFFICE				
CANTEEN				
MALE TOILETS				
FEMALE TOILETS				
WASH AREA				
MAIN STORE				
TIMBER COMPOUND				
FUEL STORE				
L.P.G. STORE				
WORK AREA: STAGE 1				
WORK AREA: STAGE 2				
WORK AREA: STAGE 3				
SUBCONTRACTORS STORE				

I THE UNDERSIGNED HAVE INSPECTED THE SITE DESCRIBED ABOVE AND UNDERTAKE TO TAKE ALL ACTIONS REQUIRED TO ELIMINATE ALL FIRE HAZARDS.

INSPECTOR OF THE SITE ....

POSITION/TITLE ....

DATE ....

Figure 105 *Form for checking site for fire hazards*

2 Install an entrance (not less than 3.000 m wide) to allow an easy access for fire-fighting equipment
3 Plan a site layout to restrict the spread of fire, and arrange for special storage of dangerous fire-feeding materials
4 Arrange for clear notices to indicate the use of any hutting and include notices to advise of any dangerous substances to be stored within, e.g. 'LPG Store' – or – 'Plant Workshop – Flammable Liquids Stored Within'
5 Ensure clear marking of fire hydrants or similar water supplies suitable for fire fighting
6 Display notices to warn of fire hazards, and include advisory notices on fire prevention

The work process and fire-fighting

The following checklist covers both site activities and workshop processes:

1 Keep a clear fire escape route and a free access for fire fighting appliances, both to and about the building
2 Maintain hydrants and other supplies for fire fighting, and keep these clearly marked
3 All hazardous materials and substances must be stored separately in clearly designated areas
4 Ensure all personnel are aware of fire escape and emergency procedures. Create a clearly marked fire point with advisory and command notices, e.g. emergency telephone number
5 Provide fire-fighting appliances in easily seen, accessible areas and arrange regular checks of these
6 Ensure all flammable liquids, and gases and their appliances are stored, used and maintained correctly
7 Provide adequate and safe heating/drying arrangements
8 Check that all electrical appliances are not overloaded, and are protected from incorrect voltage or use
9 Provide electrical appliances at the correct voltage with suitable cable transformers and fittings
10 Arrange regular clearing of combustible waste materials, e.g. rags, wood shavings, off-cuts of timber and waste flammable liquids, etc.
11 Ensure all personnel are aware of any hazardous areas marked with clear notices. Provide adequate and suitable ashtrays and/or sand buckets within permissible smoking areas, e.g. canteen, messroom, offices
12 Provide sealable metal containers for oily rags, paint rags and similar fire-inducing waste
13 Provide adequate fire-fighting aids, with particular care that extinguishers are suitable for the requirements
14 Establish, maintain and foster a good liaison with local emergency services, police and site security organizations
15 Arrange and maintain regular fire prevention checks, weekly general areas, daily to all hutting (see Figure 105)
16 Install, and make known to all operatives, a good alarm system

Fire fighting

To continue, a fire must have a supply of fuel and oxygen. The principle of fighting the fire is to eliminate either of these factors and render the flame inactive. Remember that the smoke and fumes given off by the combustion of the fire may be more hazardous than the heat. Most victims of fires are killed by suffocation caused by inhaling smoke and fumes. The following details apply to normal fire-fighting equipment, but untrained personnel must withdraw from fighting the fire if:

Hazardous materials are known to exist
The fire escape route is becoming restricted
The fire is continuing to spread, or it becomes dangerous to continue

Extinguishers containing dry powder, carbon dioxide or vaporizing liquid

These types are most suitable for electrical fires or where flammable liquids are involved. The vaporizing liquid types frequently give off dangerous fumes, and their use within confined spaces

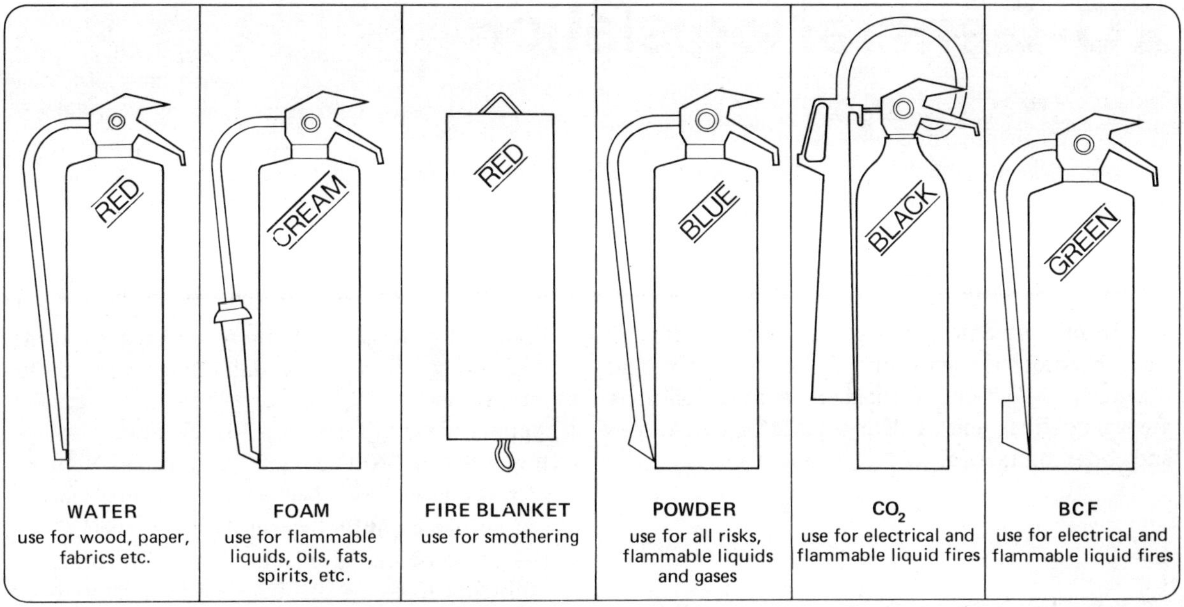

Figure 106 *Modern fire extinguishers, their colour codes and uses*

should be controlled. Any of these types of extinguisher smothers the fire by eliminating the oxygen.

The method of use is influenced by the type of fire. For flammable liquids, the user must direct the jet of the appliance at the near edge and quickly spread its contents by a cross-sweeping action, until all the flames are engulfed. When attacking an electrical fire the employee must aim the jet directly at the seat of the fire or through any opening if an encased fire.

Extinguishers containing foam

These are recommended for use on oil, grease, paint or flammable liquids. The foam ejected covers the flames and smothers the fire by stopping the supply of air. The foam should not be aimed at the base of the fire as this only forces apart the flames and induces spreading. The contents of the extinguisher must be aimed just above the fire so they fall on to it as a suffocating 'umbrella'; or the extinguisher is aimed towards the far end to build up a foam blanket from the far end of the fire.

Extinguishers containing water

These types are suitable for paper, timber and similar combustible materials, and act by cooling the fuel below its ignition temperature. The operator should aim the water on to the fuel and systematically cover the whole fire until it becomes harmless. After extinguishing the fire a continued supply of water should be sprayed on any areas that may re-ignite.

General points about extinguishers

The various types of extinguishers are colour-coded to avoid confusion or incorrect use. Figure 106 shows the various kinds of extinguisher and would be a good advisory notice to show at a fire point.

Other fire aids

Various other aids are available for use in the emergency of fire. These include an asbestos blanket to wrap around either a person or over a small fire to smother the flame. Another simple fire extinguisher for small fires is a sand bucket.

20 General legislation

The Health and Safety at Work etc. Act introduced vast changes in statutory legislation. Existing documents, Acts of Parliament and Regulations are being either amended or replaced by new laws and codes of practice.

The Health and Safety at Work etc. Act 1974

The Robens Report produced by an independent committee created by government and chaired by Lord Robens, highlighted two years of research and instigated the implementation of the Health and Safety at Work etc. Act. The Act now controls all sections of safety legislation, and has led to the reorganization and restructuring of regulations. These will be amended, adjusted, updated or replaced in accordance with this new principle Act. The Act is divided into four major groups which constitute a total of eighty-five sections and ten schedules. The details of the Act can be summarized as follows:

Health, safety and welfare at work and the control of dangerous atmospheric pollution

1 General duties of employers; self-employed; landlords; persons creating harmful substances; manufacturers of goods, machines and substances; employees and special limitations regarding safety provisions provided
2 Establishment, functions and powers of the Health and Safety Commission; Health and Safety Executive; including the power to investigate
3 Codes of practice, regulations and use of criminal proceedings in conjunction with these requirements
4 Appointment of Inspectors, their powers, Improvement Notices and Prohibition Notices, appeals, powers to deal with imminent dangers and enforcement by Inspectors
5 Details of obtaining information and restrictions of disclosure
6 Special provision relating to agriculture
7 Details of offences; prosecutions by Inspectors; onus of proof; limits of what is practicable
8 Financial provisions
9 Details of appeals; default powers; service of notices; civil liability; applicable details to the Crown; adaption to metrication as applicable to existing regulations; exclusion of domestic employees and the full meanings of work and at work

The Employment Medical Advisory Service

This deals with the involvement, responsibilities for maintaining this service including fees; other financial provisions; responsibility for accounts and supplementary factors.

Building Regulations and Amendments of Building (Scotland) Act 1959

This part deals exclusively with amendments of existing regulations; approval of plans; special provisions regarding materials for permanent buildings; civil liabilities, application to the Crown; powers to make building regulation for Inner London; interpretation of the meaning of building.

Miscellaneous and general

This small section deals with the several minor, but nonetheless important, aspects of the Act and details several amendments to existing Acts, provisions and regulations.

Factories Act 1961

This, the last and current Factories Act, deals with numerous requirements for factories, and any places of work deemed to include conditions employers offer for employment. There are fourteen parts incorporating 185 sections with seven schedules.

1 *Health (general provisions)*

The details of work areas are dealt with in this section, covering the maintenance of acceptable standards of cleanliness; overcrowding; temperatures; ventilation; lighting; drainage of floors; sanitary conveniences; certain enforcements by district councils; powers of inspectors regarding sanitation that may be remedied by district councils; powers to deal with non-complying councils; and powers cable to medical supervision.

2 *Safety (general provisions)*

This vast part of the Act deals with numerous sectors of machine parts, manufacture, use, and qualifies the requirements of training and supervising young persons. Hoists, chains, lifting tackle, floors, stairs access to and from the workplace are specified. Dangerous fumes, steam boilers, means of fire fighting, fire prevention and escape during fires are all detailed.

3 *Welfare (general provisions)*

This section details the welfare provisions that are applicable to factories for the minimum supply of drinking water; washing facilities; accommodation for clothing; sitting facilities, first aid and welfare regulations.

4 *Health, safety and welfare (special provisions and regulations)*

This part deals with special problems and provisions relating to functions and processes not usually applicable to construction. Certain details relate to removal and duration of dust or fumes; meals in certain dangerous trades; prohibition of certain female and young person employment.

5 *Notification of accidents and industrial diseases*

This covers factory situations requiring notification of accidents; notification of industrial diseases; inquests applicable to factory deaths; powers for direct formal investigations, and the duty of appointed factory doctor.

6 *Employment of women and young persons*

The factory requirements and stipulations detailed are numerous, many of which are not applicable to construction work situations.

7 *Special applications and extensions*

As the title suggests, this details special conditions applicable to many situations, of which only building operations, engineering construction and lead processes outside factories are applicable to the construction industry.

8 *Homework*

No real involvement for construction.

9 *Wages*

This deals with wages of persons employed and restrictions upon what deductions are permissible.

10 *Notices, returns, records and duties, etc.*

This indicates the need and provisions of notices, posting of abstracts of the Act and notices. Preservation of records, registers and the return of these are also detailed, as are the duties of persons employed.

11 *Administration*

Administrative details are itemized under the appointment, duties and powers of Inspectors, as are the Inspectors' entry into premises, powers to bring court proceedings and certificate of appointment. The appointment of factory doctors, their fees, etc. are also detailed with provision applicable to district councils.

12 *Offences, penalties and legal proceedings*

The sections deal with offences; fines where no set penalty exists and powers of courts to see the contravention is corrected. The proceedings regarding forgery, false entries, false declarations and various liabilities, appeals and employment of children offences are all specified.

13 *Application of the Act*

The general application and its relevance to the Crown are qualified and detailed.

14 *Interpretation and general*

A review of factors applicable to promotion of health, safety and welfare; premises to be inspected; certain regulations; certificates of birth; and a clear expression of 'factory' and general interpretations.

The Asbestos Regulations 1969

A statutory instrument that controls the problems associated with contamination created by asbestos dusts. There are six parts incorporating twenty sections:

1 *Application, interpretation and general*

The details of application and interpretation are those approved by the Chief Inspector. Asbestos is a mineral of crocidolite, amosite, chrysotile, fibrous anthophyllite or any substance containing these minerals. Further interpretations detail 'factory' and 'protective clothing' and the legal requirement to inform, in writing, the district factory inspector twenty-eight days prior to using crocidolite (blue) asbestos.

2 *Exhaust ventilation and protective equipment*

The requirements of exhaust ventilation is that suitable equipment is installed to prevent asbestos dust from contaminating any workplace. Such equipment must be inspected each seven days and thoroughly examined at fourteen-month periods by a competent person, who must confirm this with a written report. Where ventilation cannot be achieved, suitable and adequate protective garments and respiratory protective equipment must be provided and used. Apparatus used should be either new or thoroughly cleansed before use by the next person.

3 *Cleanliness of premises and plant*

This details the required cleanliness of machinery, floors, working surfaces and ledges. Legislation requires that either vacuum cleaning or a dampened cleaning process is used. Where this is not practicable, suitable and adequate protective garments must be provided if dust is created.

4 *Storage and distribution*

All small parts of asbestos must be kept in sealed containers. Waste asbestos and any dust cleared must be disposed of in sealed containers and clearly marked to confirm the contents. This applies particularly to crocidolite which must be marked, 'Blue asbestos – do not inhale dust'.

5 *Accommodation for the cleaning of protective equipment*

Where protective garments need to be supplied suitable accommodation for changing must be provided. Any garments used must be sealed in containers or sealed polythene bags marked 'Asbestos-contaminated clothing' before despatch for cleaning.

6 *Young persons*

The employment of young persons is prohibited where dust is emitted into the air unless adequate exhaust facilities are installed.

The Protection of Eyes Regulations 1974

This new regulation supersedes the Protection of Eyes Regulations 1938. Provision is made within the general requirements of the Factories Act 1961. The exact specified processes where protection must be used are detailed in Schedules 1 and 2 of the regulations. The various aspects of this statutory instrument are as follows:

1 *Citation, commencement and revocation*

The title of this regulation and those schedules listing specific details applicable are shown, including the date of coming into operation, this being April 1975.

2 *Interpretation*

Details of interpretation are given regarding 'eye protectors' – goggles, visors, spectacles and face screens. Also specified is the fixed shield, either free standing or attached to machinery, plant etc. Occasional employment is defined as an employee working at one or more processes or in a position of risk for not more than two days and not exceeding fifteen minutes in any one day. Also defined are 'factory', 'foundry', 'possession', and 'approved' for clarity of understanding.

3 *Application of regulations*

The locations applicable to these regulations include all factories and premises, places, processes or operations provided for within Part 4 of the Factories Act 1961. The Protection of Eyes Regulations is an addition to the requirements of the Factories Act and does not devalue nor substitute any part of them.

4 *Exemption certificates*

This specifies where exemption might exist for relaxation of these provisions, provided that agreement from the Chief Inspector is obtained, and a suitable certificate is issued as confirmation.

5 *Protection of persons employed in the specified processes*

The detailed obligations of the employer are laid down regarding issue of apparatus, e.g. employment as detailed in Schedule 1, Part 1 – eye protectors; employment within Schedule 1 Part 2 – a shield or fixed shield; employment with Schedule 1 Part 3 or 4 – eye protectors or a shield or sufficient fixed shields.

6 *Protection from risk although not employed on a specified process*

Where a reasonable risk exists but the employee is not actually at the dangerous or specified process, sufficient and adequate eye protection must be provided.

7 *Eye protectors – issue and availability*

The employer must provide a personal issue of protectors to those engaged on any specified process and make available an adequate supply of protectors for those occasionally employed.

8 *Replacement of eye protectors*

All equipment detailed in the preceding paragraph shall be replaced free of charge by the employer, if, at any time, an employee is left unprotected by defect, destruction, or loss of the provided protection.

9 *Eye protectors and shields – construction and marking*

All protectors issued must be manufactured to an approved specification and clearly marked to indicate the suitability and designed use of such eye protection.

10 *Fixed shields – construction, maintenance and positioning*

Fixed shields must be achieved to an approved specification, maintained in good order and protect the employee's eyes so far as practicable.

11 *Duties of employed persons*

Where the Regulations indicate eye protection is needed, all employees must take reasonable care of the protectors, report any defect or loss of protectors and take reasonable care to eliminate risk whilst engaged at a specified process.

Specified processes applicable to construction

The following work requires the use of eye protectors:

1 Blasting or erosion of concrete by means of shot or abrasive materials or compressed air
2 Cleaning of buildings by the above processes
3 Cleaning by high-pressure water jets
4 Striking masonry nails by hand-tool or powered hand-tool
5 Any work with a hand-held cartridge-operated tool
6 Chipping of metal or cutting, knocking metal fixings (bolts, rivets, etc.) with hand-tool or powered hand-tool
7 Chipping, scurfing of paint, etc. or other corrosion by hand-tool or powered hand-tool
8 Any high-speed metal-cutting saw or abrasive cutting-off wheel
9 Any activities applicable to plant used in conjunction with acid and similar dangerous substances
10 Any handling of open vessels containing acids or similar dangerous substances
11 Driving of any bolts, pins, collars, etc.
12 Injection by pressure of any liquids or solution into buildings
13 The breaking, cutting, etc. by hand-tool or powered hand-tool of many building materials, e.g. glass, concrete, plastics, stone, plaster, brickwork, blockwork bricks, tiles and blocks (except those made of wood)
14 Cleaning by compressed air

The following work requires the use of approved shields or fixed shields:

1 The welding of metals
2 The cutting, boring or cleaning of metals where the apparatus uses air, oxygen, or any flammable gases
3 Any process where light amplification or dangers of radiation exist
4 Turning or dressing of an abrasive wheel
5 Dry grinding of materials, where fragments may be thrown off

In the following situations persons are not employed on a specified process but may be exposed to risk:

1 Any chipping of metal, knocking, cutting, etc. of metal fixtures (bolts, rivets, etc.)
2 Any process involving an electric arc
3 Any process where light amplification or dangers of radiation exist

A detailed review for complete terminology should be sought in the regulation itself where the lists given do not fully itemize all particulars.

The Construction (General Provisions) Regulations 1961

These regulations detail controls on construction processes to promote a good level of general safety standards. The documentation is far-reaching, demanding, and within twelve parts contains fifty-seven sections. The following information is included:

1 *Application and interpretation*

This section explains the details of title, application and obligations of the document and defines the meaning of several terms, e.g. 'plant or equipment', 'gear machine', 'apparatus or appliance', 'scaffold board', etc. (Scaffold is deemed to be any temporary working platform and its supports.)

2 *Supervision of safe conduct of work*

This details the obligations applicable to the appointment of a safety supervisor where any company employs more than twenty persons, whether or not they are on one site. The appointment, confirmed in writing, is to uphold safety standards for a contractor or group of contractors.

3 *Safety of working places and means of access*

This section, which explains the need of safe working provisions, is now revoked and incorporated within the Construction (Working Places) Regulations 1966 (page 107).

4 *Excavations, shafts and tunnels*

Where an excavation, tunnel or shaft exceeds 1.219 m and a danger of material or soil collapse exists, there must be adequate support to the sides and protection to all operatives. This does not apply if the sides have been battered back, or where specialist tunnelling is to be done.

An inspection, without a written record, must be made in these circumstances at least every day when work is active and at the commencement of every working shift at the workface, where the excavation exceeds 1.980 m.

A thorough examination, recorded within the official register, must be detailed on Form 91 Part 1 Section B (Figure 7, page 41) at least every seven days, or after use of any explosive charge, or after any collapse of material or earth.

All work of timbering must be completed with good, sound materials and equipment which has been inspected before use.

Wherever a risk of flooding or similar hazard exists within a tunnel, a good means of exit must be provided, usually by way of a ladder.

Where existing buildings are likely to be affected by the excavation activity, a suitable means of shoring or propping must be provided, except within specialist tunnelling activities.

Where an excavation exceeds 1.980 m and is adjacent to any walking area for other operatives, adequate guards or barriers must be provided.

Provision must be provided to eliminate risk of plant or machines falling into any excavation.

5 *Cofferdams and caissons*

All activities must be properly completed and maintained, after completion, with ladders or similar means of escape in case of flooding. All work must be supervised by a competent person and an unrecorded inspection completed before every shift. A recorded thorough examination must be carried out, with Form 91 Part I Section B being completed, every seven days or where explosive charges have been used or whenever damage has occurred.

6 *Explosives*

Explosives may only be used by trained experienced persons, who must give adequate warning of any charge to be made.

7 *Dangerous or unhealthy atmospheres*

If dangerous fumes or dust are expected, a good means of ventilation or personal respirators must be provided.

Where these unhealthy atmospheres are expected in any excavations, ventilation or breathing aids must be provided where poisonous gases are found, adequate tests must be completed and waiting periods observed.

8 *Work on or adjacent to water*

Where this problem exists, a well-constructed and adequately maintained craft must be used by an experienced boatman. If the vessel has to carry more than twelve persons, it must be suitably tested.

9 *Transport*

All railway tracks must be laid on firm foundations with suitable buffers at track ends. A good clearance around any track must be allowed to avoid persons being crushed.

Any locomotive or similar mechanically propelled vehicle must be driven by a person over eighteen years old unless that person is under training to drive, and is personally supervised.

No one should be allowed to ride on a moving vehicle unless it is specifically a passenger transporter. Adequate measures must be taken to avoid vehicles falling into any excavation or over the edges of embankments.

10 *Demolition*

Any employer engaged upon demolition must appoint an experienced supervisor, who must arrange all safety work activities and timing of operations. Where the demolition contractor is self-employed, and working alone, he must discuss, advise and warn all other site contractors or employees.

Adequate care must be taken to avoid fire, flood or explosion. Regular checks to all services must be arranged by a competent supervisor.

All working activities must be arranged by a trained competent supervisor. The possibility of floor overloading, sudden collapse, excessive loading of parts of the structure, in general, any dangerous practice must be eliminated.

11 *Miscellaneous*

All dangerous moving parts of machines must be guarded, and all new equipment must be provided to the highest safety standard by all manufacturers. All electrical apparatus must be installed, maintained and checked by a competent person who must have adequate electrical knowledge and experience.

Where any employee is to be exposed to overhead dangers, adequate head protection, or overhead protection must be provided.

All employees must be protected as far as reasonably practicable from projecting nails, poorly stacked materials, inadequate lighting, temporary supports to buildings or parts of buildings, and excessive personal lifting. All records must be prepared, maintained and available as directed by the factory inspectorate or as directed by each statutory document.

The Construction (Lifting Operations) Regulations 1961

Another complex piece of legislation, applicable to the lifting activities of construction processes. Within the seven parts, fifty sections contain the following requirements:

1 *Application and interpretation*

This explains where the Regulations apply, and stresses that the employer of those using any plant is responsible for its compliance to all legal requirements. Every employer must also check that guarding, lifting and work procedures comply with all Regulations.

Some typical interpretations are: a hoist is a machine for lifting men or materials; a mobile crane is one that is on road wheels and not track mounted; safe work loads (SWL) are those shown on the appliance for which they have been tested and suitable certificates issued.

2 *Exemptions*

Certain exemptions can be obtained from HM Chief Inspector of Factories, but only in that minority of situations in which compliance is not practicable. A typical exemption is for lifting gear fitted to materials or plant before delivery; it need not comply with the Construction Regulations, if strong enough and used only for the site delivery.

3 *Lifting appliances*

All appliances must be strong, well made, maintained in good order and inspected each week, with a record of inspections kept in the register (Form 91 Part I, Section C as shown in Figure 8, page 43). All foundations and temporary supports for appliances must be adequate and a clearance of 0.610 m around travelling or slewing appliances must be achieved wherever practicable. (See Figure 107 for a typical example.)

A good means of access must be available to all parts requiring inspection or maintenance located where the operatives might fall a height of 1.980 m.

Any pole or support for a lifting device must be strong enough to resist movement or failure of the appliance.

Details applicable to cranes are shown. All crane appliances must be securely fixed and correctly balanced with a ballast. Records of tests and inspections must be recorded.

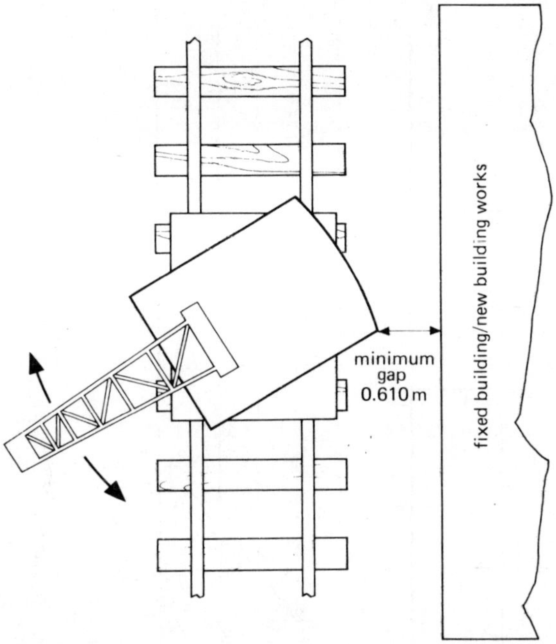

Figure 107 *Clearances around moving plant and machinery. The minimum gap is 0.610 . Often the clearance varies as the machine swings or moves: the gap must be checked when it is at its smallest, as here*

All drivers and signallers of lifting appliances must be competent and over eighteen years of age unless under training. All signals must be clear, concise and easily understood.

Testing must be completed before lifting appliances are used on the construction site. Tests and examinations must also be completed after a major adjustment. Examinations must also be carried out at fourteen-month intervals, and recorded on Form 91 Part 2 Section G (page 42).

4 *Chains, ropes and lifting gear*

All chains, ropes, etc. must be properly made, with a certificate confirming the test and examination, together with the safe working load. Any wire rope with more than one in twenty strands defective within a length of ten wire diameters must not be used.

All hooks must be fitted with a safety device to prevent the load from slipping off the hook. Wires joined by tied knots or bolts, passing through links, must not be used.

All ropes, wires, etc. must be examined every six months with a record in Form 91 Part 2 (see page 45).

5 *Special provision as to hoists*

Provision must be made to prevent anyone being injured by hoists at ground level, or at each floor access, by incorporating a suitable protective gate.

Only one means of control is permissible. Where the driver/operator cannot see clearly, a signalling system must be operated.

Hoists must have over-run stops at the top, and an automatic brake system when the control rope is released.

Materials hoists must be marked with a safe working load indicator and a notice prohibiting passengers. Hoists for passengers must be marked with a safe working load and the maximum permissible number of passengers.

All hoists must be tested and examined before use on site, and examined each six months thereafter, with a record made of these examinations on Form 91. In addition, all hoists must be inspected weekly and these recorded on Form 91 Part 1 Section C (Figure 8, page 43).

6 *Carriage of persons and secureness of loads*

No passenger hoist must allow its occupants to fall out, be hit by any falling object, or trapped between cage and any structure. The gates must be interlocking to prevent opening, other than at prepared landing positions.

All loads must be secured to prevent slipping, falling or any similar danger. Loose materials must be in containers or wheelbarrows which need to be secured or prevented from moving or tipping.

7 *Keeping of records*

Where the work is to last for more than six weeks Form 91 must be completed in respect of weekly inspections, anchorage and stability, automatic indicators, and passenger hoist tests.

Factories Act 1961

CONSTRUCTION (HEALTH AND WELFARE) REGULATIONS 1966

F 2202

Certificate of shared welfare arrangements made under regulation 4

Form approved by H.M. Chief Inspector of Factories

PART A

(To be completed by the contractor providing facilities and handed to the contractor for whom facilities are provided)

Name or title of Employer or Contractor providing the facilities

____

Address of Site

____

____

Name of Employer or Contractor for whom facilities are provided

	Facilities provided						
	First-aid boxes (Regs. 5(1), 10(2) and Schedule.)	Trained first-aider (Regs. 5(2), 7 and 10(2).)	Ambulance arrangements (Regs. 8 and 10(2).)	First-aid room (Regs. 9 and 10(2).)	Shelters and accommodation for clothing and taking meals. (Reg. 11.)	Washing facilities (Reg. 12.)	Sanitary conveniences (Reg. 13.)
Whether facilities provided (Yes/No)							
Date arrangements began							
Date arrangements ended							

SPECIMEN

Note

This certificate must be kept by the contractor for whom facilities are provided. It must be kept on the site or at his office and must be available for inspection by H.M. Inspectors of Factories or any employee who is affected by the arrangements.

(Signed)______________________

For and on behalf of ______________________

(Name of Employer or Contractor providing facilities) ______________________

Date______________________

Figure 108 *Certificate of shared welfare arrangements made under regulation 4, Construction (Health and Welfare) Regulations 1966. A copy is retained in the register*

The Construction (Health and Welfare) Regulations 1966

Further details of the maintenance of safety standards are set out in this Regulation, and can be summarized as follows:

1 *Application, interpretation and obligations*

These regulations apply to all construction or maintenance work, subject to special exemptions granted by the Chief Inspector of Factories.

The applications specify a certificate in first aid as one being issued by an approved organization or society to anyone over fifteen years old and having been issued within three years.

Suitable training organizations or societies are deemed the St John Ambulance Association, St Andrew Ambulance Association, British Red Cross or suitable local authority training schemes. The obligations on every employer are to provide facilities laid down by these Regulations, but that certain facilities can be shared by mutual agreement in writing. Shared facilities include first aid boxes, ambulances, first aid rooms, shelter messrooms, washing and toilet facilities. Where facilities are shared, the provider assumes responsibility for all who use these facilities. A register must be kept and a certificate issued to each firm sharing such facilities (in Figure 108).

2 *Provision of first aid boxes*

Where anyone employs more than five persons on a site adequately stocked first aid boxes must be provided. A clear notice displaying 'First aid' and the name of the first aid attendant must be shown. Several boxes must be provided for scattered working. If the contractors employees total more than fifty, the first aid attendant must hold a valid first aid certificate.

The contents of the first aid box must be regularly checked and properly maintained.

3 *Ambulances and first aid rooms*

If he employs more than twenty-five persons, an employer must contact the local ambulance authority, provide at least one stretcher and appoint someone responsible for contact throughout the working day.

On any site where the total people employed exceeds 250, any contractor who employs forty or more must provide a convenient first aid room containing the equipment specified by this regulation. (Three construction office workers are calculated as one site worker for these purposes.)

4 *Shelters for clothing and taking meals*

On every site there must be provided suitable hutting as protection from bad weather, and in which to leave personal or protective clothing and to take meals. Seating facilities must be available. Where more than five men are employed, the shelters should be heated. If more than ten persons are employed, provision for heating food should be made, unless there are other arrangements.

5 *Washing facilities*

On every site where employment exceeds four hours, some form of washing facilities must be provided. Where the contractor employs more than twenty men or the work will exceed six weeks' duration, the facilities must include basins (or buckets), soap and towels (or driers) with hot and cold water. Where the contractor employs more than one hundred men or the work will exceed twelve months the facilities must include four basins or buckets (and one more for each further thirty-five men). Where dangerous substances are to be used, e.g. lead, the employees involved must have one basin (or bucket) for each five persons, with soap, towels, water and nailbrushes supplied.

6 *Sanitary conveniences*

These must be provided in clean, accessible, covered and well-ventilated conditions. One convenience is required for each twenty-five persons.

7 *Protective clothing*

Every employer must provide suitable and adequate protective clothing for working in snow, sleet, rain, etc.

The Construction (Working Places) Regulations 1966

This statutory instrument controls all working places, but concentrates mainly on scaffolding and similar temporary structures. The main parts of this document relate to scaffolding (pages 102-17). This summary deals with relevant details other than scaffolding.

1 *Application and interpretation*

The regulations include site clearance, demolition, foundations, the structure of all ancillary work applicable to building or construction operations.

The obligations upon every employer to provide a safe workplace for each of his employees are detailed. Special arrangements must be made by employers who, by their actions, may endanger employees other than their own, e.g. where using explosives, cranes, etc.

Interpretations include definition of various scaffold types.

2 *Exemptions*

Certain exemptions apply, e.g. where compliance is not practicable, or where safety can be main tained without compliance.

3 *Safety of working places and access and egress*

The whole of this section is fully reviewed on pages 102-24.

4 *Keeping of records*

Reports applicable to the inspection of scaffolds either each week, after bad weather or after structural alterations, must be made in the Scaffold Register Form 91. This should be available on site or at the contractors' office where work will last for more than six weeks.

21 Carpentry and joinery safety

Many details in previous chapters apply to individual sections of construction. This chapter includes details relevant only to carpentry, joinery and allied work.

Mechanical equipment

Wherever a woodwork machine is established, whether on site or in a workshop, it must comply completely with the legislation applicable to its use (see pages 179-83). In practical terms, however, the following need careful consideration:

Location/position

There must be adequate space around each machine to allow its full function. This has been explained on pages 78-9 and is particularly relevant to wood-machining. This requirement is particularly abused in site operations where machines in obscure positions create danger.

Use

Where a machine has been designed, manufactured and installed for a particular job, it must be employed solely in that capacity. Improvisation and short cuts are not acceptable. Both employees and employer must discourage such dangerous practices.

Routers

Of the range of powered hand-tools reviewed on pages 134-41, this particular appliance is used only by joiners. The router has a motor which drives a central chuck. A cutter (or bit) is secured to the chuck, and rotates at high speed. This is fed along the workpiece. A wide selection of cutters are

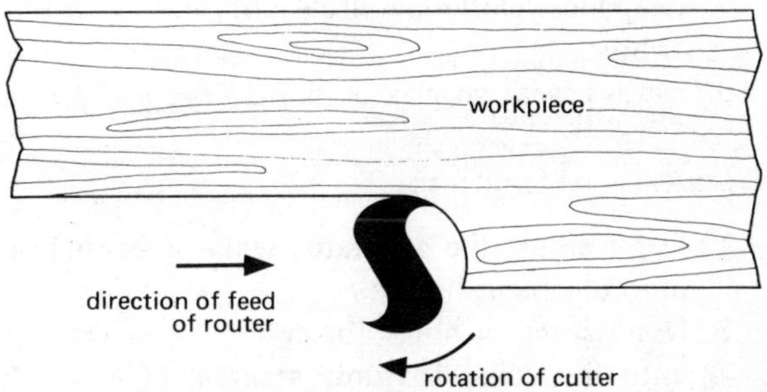

Figure 109 *Portable router cutter*

available, along with jigs and templates for operations such as dovetailing. Routers are usually double insulated, indicated by the international symbol ⧈. This means an external cover to the tool gives a second, independent insulation barrier. All details of wiring, connections, plugs, etc. are as on page 141. The chuck is equipped with a 0.012 m collet which is usually tightened by a wrench (but without excessive torque or pressure). In certain modern designs no wrench is needed to change the bit. Various attachments and jigs are required for different operations, and the recommendations in manufacturers' literature must be observed.

Routers – good practices

1 Always use correct accessories specifically designed for high-speed routers
2 Never adjust the appliance without removing the plug from its socket
3 Switch on the motor and allow it to reach full revs before commencing the work
4 Retain a firm grip of the router when switching on to resist the starting torque of the motor.

Always make sure the workpiece is firmly held before commencing work

5 Pull the router towards you with the open side of the cantilevered base facing you
6 Use the cutters in the correct way. The motor rotates clockwise when viewed from the top. (See also Figure 109)
7 Use only sharp cutters
8 Make sure that the ventilation passages are clear, the air holes free from build-up of dust
9 Keep loose clothing well clear of rotating cutters or bits
10 Always wear goggles or similar eye protection

Routers – bad practices

1 Don't adjust the apparatus while it is plugged in to the mains supply
2 Don't force or abuse the collet. Never operate until the collet is firmly secured. (The shank must be in by at least 0.012 m)
3 Never lay down the router with the bit rotating
4 Never work slowly across the workpiece – the cutter will overheat and also burn the material
5 Don't try to speed up the feed
6 Where the work is away from the material edge don't plunge in. Allow a steady entry before cutting commences
7 Don't attempt a deep cut in one pass. Achieve this gradually in stages
8 Make sure the cutter is free from the workpiece before turning on
9 Never operate the router in damp or wet conditions
10 Never operate the appliance with damaged or defective wiring. Keep all cables well clear of the cutters whilst working

Materials

The basic materials of the wood trades are relatively safe, provided storage is correct and properly maintained. Stacks of materials must not exceed 2.000 m in height unless suitably restrained within racks. Sheet materials must be stored either flat or within a rack if vertical stacking is acceptable. The following materials are considered to present hazards to carpenters and/or joiners.

Adhesives

Adhesives are composed of numerous chemicals and substances which are harmful to skin and/or internal body organs. Where skin diseases may be caused by handling adhesives, suitable hand protection must be used – gloves or a hand barrier cream. Good personal hygiene is also necessary before food breaks or at the end of the working shift. These are predominantly the employee's responsibility. There are also obligations on the manufacturer to provide advisory or warning instructions and on the employer to inform his employees of all dangers relating to the adhesives. Dangers can occur by inhaling toxic fumes from adhesives; mouth masks must then be worn, particularly with petroleum-based adhesives. Smoking is also a serious hazard with certain adhesives, and a special care is necessary.

Preservatives/insecticides

Numerous types are available, each containing different ingredients for their function. Almost all chemicals used for these timber treatments are harmful, and most are poisonous. The poison is more active in the liquid form of such preservatives, but care must also be taken when using treated timber. Hand protection by gloves or barrier cream is necessary, along with good hygiene habits. Cutting treated timber by machine or powered hand-tool creates dust which is harmful if inhaled. Mouth masks must be worn along with goggles to avoid infection affecting the eyes. Any employee who is treating the timber needs similar protection.

Lead paints

In good carpentry and joinery paint is used in prefabrication and assembly activities. (Pages 194-5 provides guidance on procedures.)

Equipment

The full range of equipment used by this trade is enormous. As always, good advice is, 'Take precautions, so far as reasonably practicable.' As well as the hazards mentioned in other chapters, the following are of particular importance.

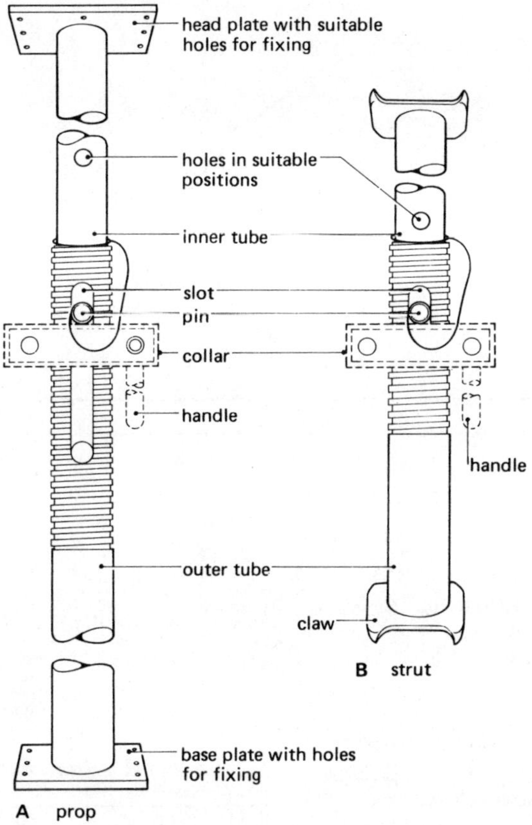

Figure 110 *Adjustable steel prop (A) and strut (B)*

Adjustable props (Figure 110)

Whenever these are used as temporary supports they must be secured at the top and bottom, to head and sole plates respectively. This avoids collapse, and side movement. The supporting pin must always be the one provided for the job – no improvising (e.g. with steel nails) should be allowed.

Column cramps

These should be supported by nails to prevent their slipping down. The ends should be painted white or a bright colour to warn all employees.

Forklift trucks

These can be used advantageously for mechanical lifting (pages 187-8).

Legislation

The main legal control on wood trades is the Woodworking Machines Regulations 1974. Other statutory instruments relevant to carpenters and joiners have been considered in other chapters.

The Woodworking Machines Regulations 1974

This legal document superseded all previous controls and was introduced in November 1974, except for Regulation 41, which became law in March 1976. The machines affected by this law are:

Any sawing machine with one or more blades – circular saws
Grooving machines – radial arm trencher
Any sawing machine with a continuous band – band saw
Chain saw machines – cutting within log mill
Mortising machines – chisel or chain mortisers
Planing machines – table and thicknessing planers
Vertical spindle moulder – including high speed routers
Multi-cutter moulding machine – four, six or eight cutters
Tenoning machine – single – or double-ended tenoner
Trenching machines
Automatic or semi-automatic lathes
Boring machines

These Regulations apply to static machinery and portable powered hand tools. The Regulations specify a clear interpretation of the detailed meanings applicable to the machines.

Machine guards and adjustments

All guards must be designed to enclose cutters, so far as practicable, depending on the work being done. All guards must be secure and of substantial construction. No adjustment to a guard or machine is permissible while the machine is in motion.

Use and maintenance of guards

Guards provided must at all times be correctly used. Where the nature of the work renders this impracticable, reasonable precautions must be taken. There are eight situations in which no exception is permitted.

Working space/floors

The working space must be adequate and unobstructed. Details vary according to the work process and component to be manufactured. All floor areas around any machine must be level, non-slip and free from offcuts and obstructions (see Figure 31).

Temperature

The minimum temperature allowable is 13°C for workshop and 10°C for sawmills. Special provisions for heating must not increase the fire risk.

Training

Woodworking machines may only be operated by persons over eighteen. Sufficient training must be given or adequate supervision maintained.

Duties of employee

The obligation on employee is to take reasonable care, using all guards, spikes, push sticks, push blocks and any other device required.

Circular sawing machines

These are defined in the Regulations as a bench of fixed or rolling table, having a spindle below the table to which a blade can be secured for dividing materials. The following requirements apply:

1 The blade below the bench must be enclosed as far as practicable
2 There must be a riving knife, securely fixed, behind the line of the saw blade (Figure 112)
3 The maximum gap between the top edge of squared stock material and the lower edge of the crown guard must not exceed 0.012 m when being hand fed to the machine (Figure 111)
4 The side flanges of the crown guard must protect both sides at least to cover the gullet of the saw blade (also known as the teeth root). Any extension guards affixed to the crown guard may be single sided

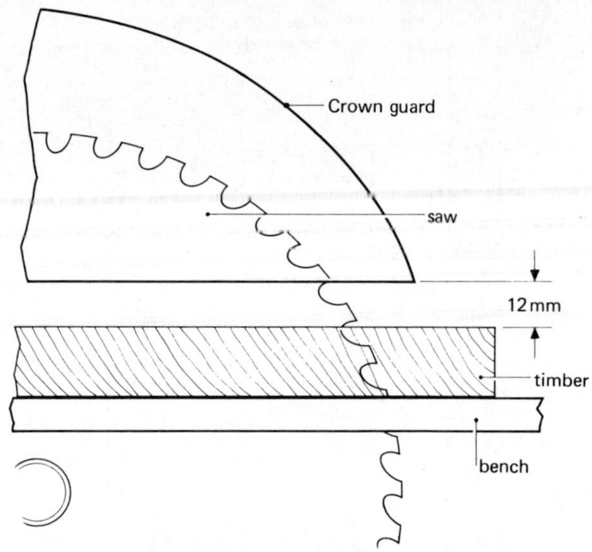

Figure 111 *Crown guard on a circular saw (Regulation 16 (3))*

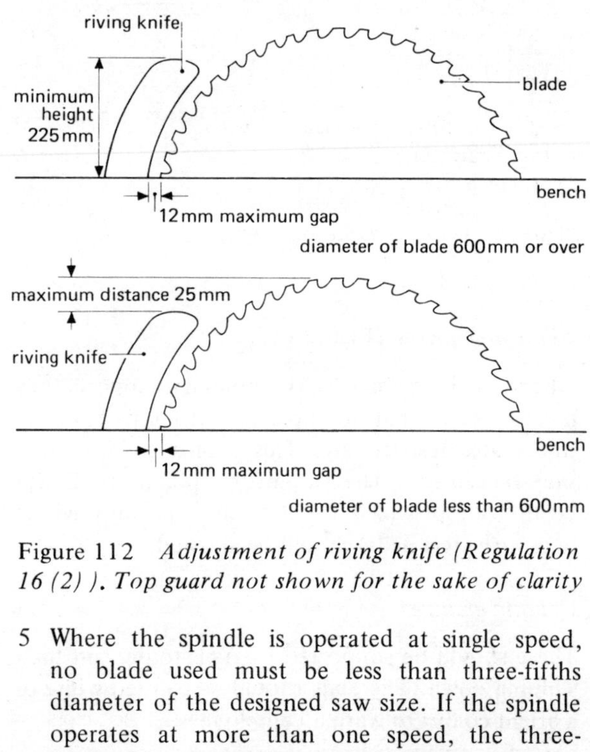

Figure 112 *Adjustment of riving knife (Regulation 16 (2)). Top guard not shown for the sake of clarity*

5 Where the spindle is operated at single speed, no blade used must be less than three-fifths diameter of the designed saw size. If the spindle operates at more than one speed, the three-fifths diameter applies to the designed size of the largest blade to the fastest spindle. A notice specifying the smallest permissible blade must be displayed

6 Rebates, tenons, moulds, grooves cannot be cut on a circular saw unless it can be adequately guarded and/or the blade projects through the upper surface of the material. However, in most cases this is not practicable
7 Push sticks designed to keep the operators hand at least 0.300 m clear of the cutting edge, must be provided at the machine and used by all operators
8 There must be a table, either static or rolling, of at least 1.200 m, behind the up-running part of the saw blade, for the full width of the saw bench. Exemptions apply to saw benches with rolling or travelling table and to any moveable machines accommodating a blade up to 0.450 m diameter

Multiple rip sawing machines

Every multiple rip saw or straight line edging machine must be fitted with in-feed pressure rollers or similar device to avoid accidental ejection of materials. These anti-throwback fingers and side guards to all blades above the table are required to give additional protection to operators.

Narrow band sawing machines

These are defined as having a blade width, not exceeding 0.050 m in a continuous band, or strip running in a vertical direction. They do not include log band saws or band resawing.

All parts of the band saw must be completely guarded except its cutting area between the top wheel and machine table. The guarding to the upper wheel must be complete to back, front and flanged as close as practicable to the side of the wheel.

Adjustment to the friction disc, or rollers, must be kept as close as practicable to the work bench.

Planing machines

These are defined as a machine for surfacing, thicknessing or a combination of these, but excluding multi-cutters with two or more cutter spindles. The following requirements must be followed:

1 Cutting of rebates, tenons and moulds is acceptable provided adequate guarding is maintained

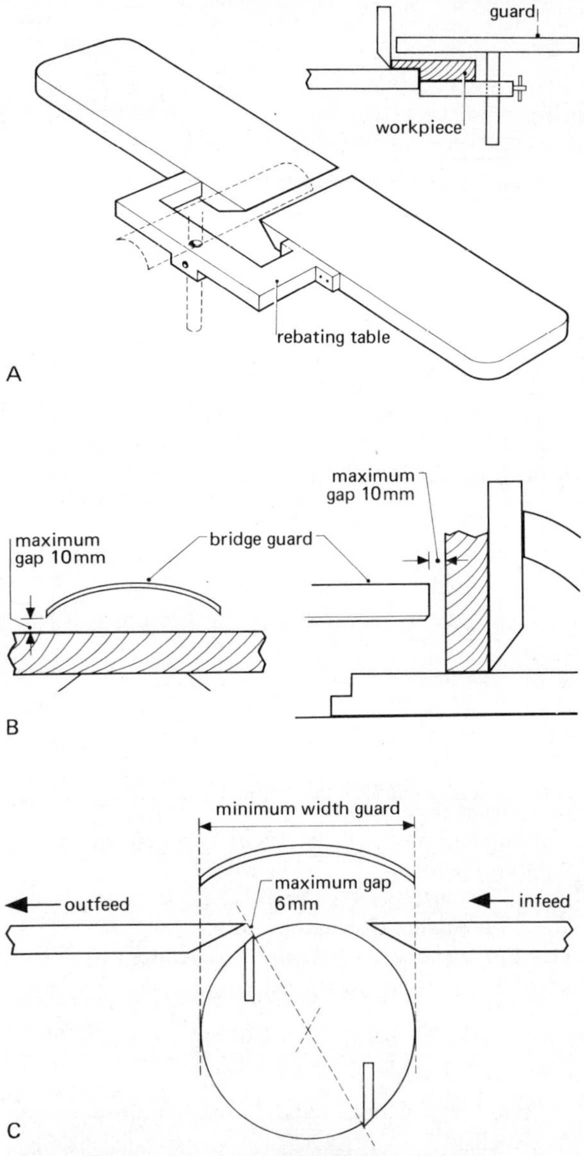

Figure 113 *Using a surface planer.*
A – Where it is not possible to use the planer for rebating because the bridge guard has to be taken off, the planer must be fixed with a suitable additional rebate table
B – The setting of the bridge guard should give a maximum gap of 10 mm, whether the workpiece is to be surfaced or edged
C – The contours of the bridge guard should be similar to those of the perimeter line of the cutter block. The maximum gap between the cutting edge and the table is 6 mm

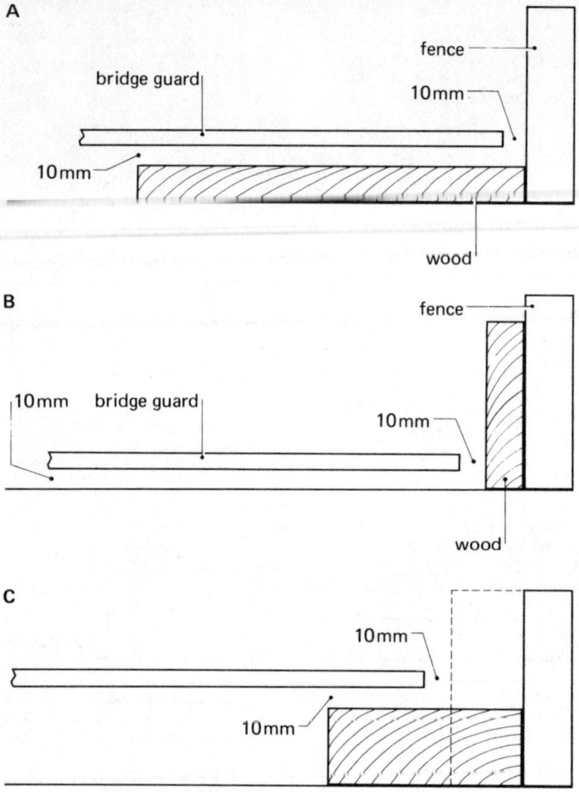

Figure 114 *Required positions of bridge guard on a planing machine.*
A – For 'flatting' (Regulation 27 (2))
B – For 'edging' (Regulation 27 (3))
C – For 'flatting' and 'edging' carried out one after the other (Regulation 27 (4))

2 Hand-fed planers must have cylindrical cutter blocks
3 The gap between cutters and front of the delivery table must not exceed 0.006 m
4 All planers which are not mechanically fed must have a strong and rigid bridge guard, not less in length than the cutters, and in width not less than the diameter of the cutter blocks. The bridge guard must be adjustable and retain safety so far as practicable. (See Figures 113 and 114 for the acceptable adjustments for planers)
5 A suitable guard must be provided over the cutter block behind the fence of the machine

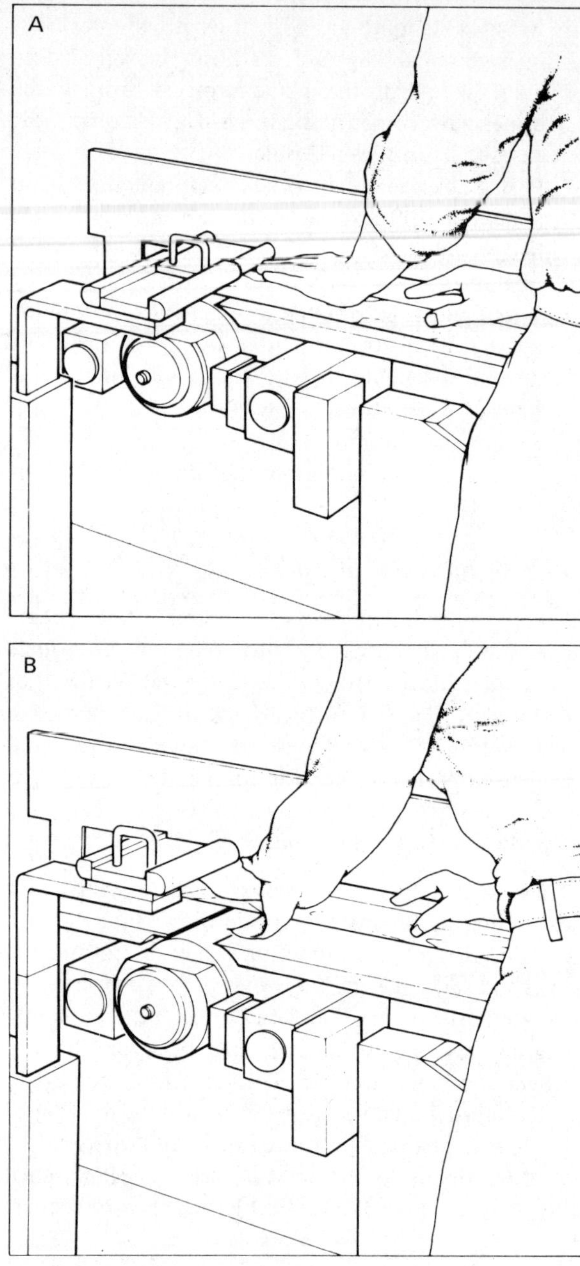

Figure 115 *Use an overhead planer correctly.*
A – Right
B – Wrong. The operator's clothing is loose and liable to be caught in the rotating blades. Too much of the blade is exposed and the vertical fence is too far away. The bridge guard is set too high to be effective, allowing easy injury to the hand exposed to the cutter blades

6 Where it is impracticable to achieve the work process with the bridge guard, a suitable push block must be used
7 Where combination planers are used for thicknessing, a suitable guard must cover the cutter block. A suitable exhaust hood satisfies this requirement where applicable
8 Where combination planers are used for thicknessing, there must be sectional-fed rollers, or anti-kickback fingers, at the operator's end of the machine. If this cannot be achieved only single pieces can be fed through, and a clear notice stating this must be shown

Vertical spindle moulding machines

This class of machine includes high-speed routers and must achieve the following requirements:

1 All cutters must be the correct thickness for cutter block and be secured to prevent them from being detached
2 The gap between fences must be reduced as far as practicable for straight-line work. This is best achieved by a false fence spanning the gap, and allowing only cutters to project
3 Where guarding is complete and the cutters may still be exposed, suitable jigs or holders must be provided and used. This applies mainly to stopped work or curved work
4 All cutters and blocks must be guarded completely, particularly to the front and top of the spindle. A hood above and around the spindle is required to achieve this
5 If the process is begun part-way along the material, a suitable back stop must be provided to restrain the work at commencement of cutting
6 Push sticks must be provided and used as an aid to safety
7 If the machine has two speeds, it must be arranged to run at the slower speed before the faster one

Extraction equipment

The main part of this section came into force in May 1976 and requires the following.

Suitable exhaust systems must be used to extract chips and other particles of material from specified machines. Exceptions apply to high-speed routers that blow away chips or any vertical spindle moulder or tenoning machine used for less than six hours per week.

Lighting

Adequate lighting must be provided and be suitable for machines in use. Glare or light must not shine into the eyes of the operator (see page 80).

Noise

Where any person is exposed continuously to a noise level exceeding 90 decibel (dbA) for eight hours per day, the noise level must be reduced, or ear protectors provided to all persons in the vicinity. This is best achieved by enclosing the noisy machine within a booth. Where the noise level is excessive for short duration or at a particular place, ear protection must be provided to persons affected (see also pages 69-70).

General factors

The carpenter and joiner is likely to meet specific hazards which require special considerations. Some of the more dangerous examples are reviewed below.

Working on open joists

Open joists are a working place which can be a danger area if not respected. Joists must always be secured by lathe or battens temporarily, until final fixing is achieved. Walking directly on the joists should be avoided, particularly while lifting. Boarded walkways should be provided for any foot traffic to all joists.

Glazing

This area of work is done by various trades according to regional habit. Plumbers, painters or carpenters have all been known to do this job in addition to sub-contractors. Regardless of trade, the employee should observe the following:

1 Always carry more than one pane to avoid the whipping effect of glass

2 Ensure the path is clear of obstructions
3 Be aware of other people, particularly behind you if you are stopping
4 Handle glass with care. Small panes should be carried underarm, and larger panes with hands beneath and to the front edge of the pane
5 Carry glass close to the body and use protective rag or gloves on your hands
6 Store glass in a dry place, on edge, with the lower edges protected by blocks, and the back with soft padding
7 Do not store glass flat on the floor
8 Fix glass immediately. Never leave glass about on site – it should be either stored or fixed
9 Ensure a good fixture of glass to avoid future accidents to employees or occupants

Projecting nails

Within the context of Regulation 48 of the Construction (General Provisions) Regulations 1961 no nails should be left protruding through any material. The carpenter on site is the most predominant user of nails, and he must take care to avoid this danger. Protruding nails must always be withdrawn or at least bent over to make them harmless. An additional safeguard is that all site employees wear good footwear, ideally safety boots with flexible steel soles (see pages 99-100).

22 Bricklaying safety

Bricklaying and associated activities are hazardous in many ways, although this chapter is relatively brief. Many safety practices applicable to bricklaying are to be found in the chapters covering scaffolding, trench supports, site transport and site dangers. All bricklayers should also read these other chapters to secure details of how good safety procedures can be achieved.

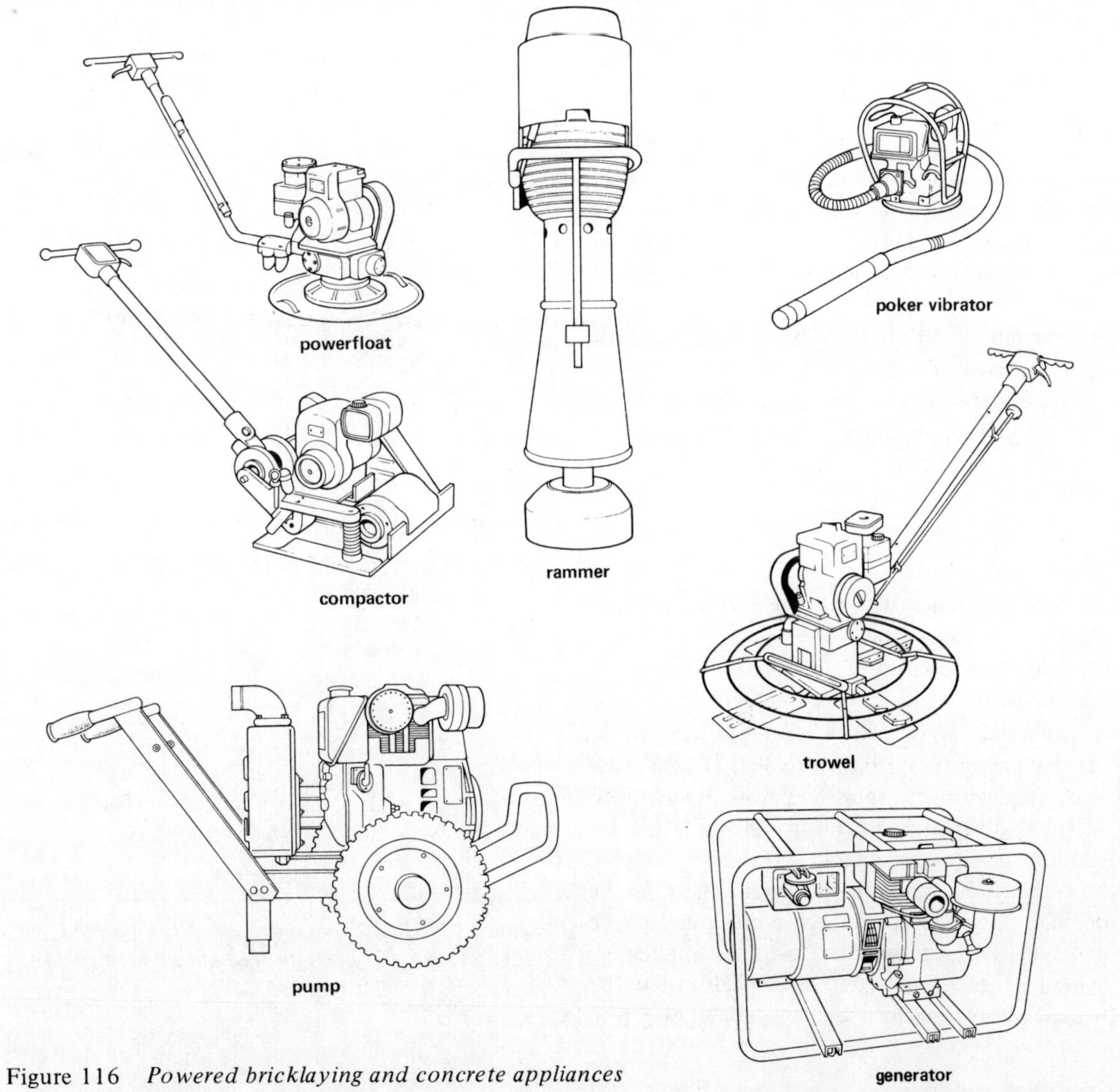

Figure 116 *Powered bricklaying and concrete appliances*

Mechanical equipment

Concrete breakers

These small tools, powered by electricity or compressed air, are used for concrete-, brick- or stone-breaking work. Chisels slotted into the nozzle end are activated by the motor, which forces the tool into the material. The normal details of connection, installations, power, couplings, cables, hoses etc. are discussed on pages 134-41, 146-8. Body protection required consists of goggles and steel toe-capped boots. The vibrations caused by the machine can lead to nerve deficiency. Gloves are required to reduce this problem, and a rota system of employees will prevent any one person being subjected to excessive periods of vibration.

Disc cutters

These are powered by electricity, or in certain cases compressed air. It is a hand-held power tool which requires the normal connection and power procedures (pages 134-41 and see also pages 146-8). Goggles are very important to achieve protection from dust and similar particle hazard. Heavy duty, steel toe-capped boots must be worn to prevent any loose parts of a cutter from damaging the user's feet. The employee must always wear a mouth mask or respirator, according to the severity of the dust produced.

Clipper saws

These static machines are used for cutting masonry products. Hand disc cutters are used to cut out work *in situ* and the appliance passes over the workpiece. Clipper saws are, however, secured within a workshop or similar permanent location and have workpieces clamped in for cutting purposes. The machine must comply with the Abrasive Wheels Regulations 1970 (page 188). The operator must observe his obligation to take reasonable care in complying with the basic machine requirements – he must be competent, trained, etc. Goggles must always be worn and ear protectors where applicable. A visual check of abrasive wheels may be sufficient. A tap with a light, non-metallic tool

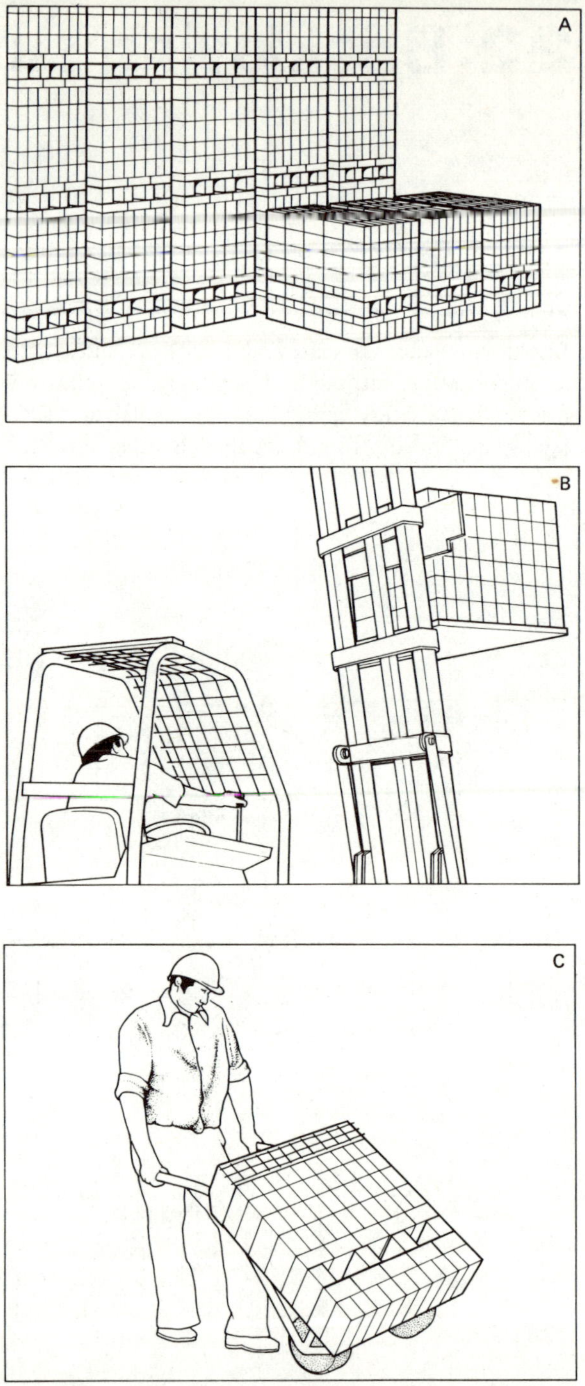

Figure 117 *Brick transport. Bricks in banded stacks (A) can be safely moved about with forklift trucks (B). Smaller banded stacks are moved with trolleys (C)*

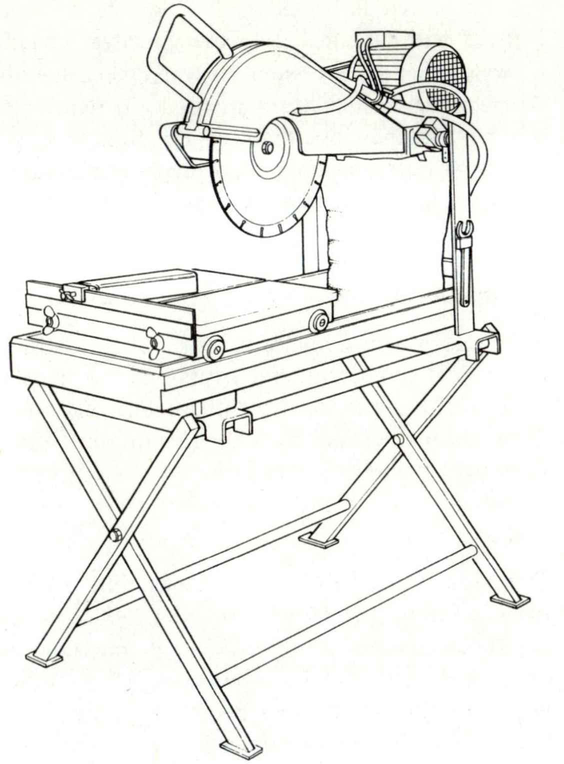

Figure 118 *Electric masonry cutter*

gives either a clear ring, or a dead sound. The latter confirms a cracked wheel which must not be used. Careful wheel storage should be provided to eliminate rolling or bumping. Cradles or racks are most suitable as they provide individual accommodation to each wheel.

The workpiece must not be offered towards the cutter at its corner. The blade must first be applied to the flat surface and worked towards the edge (Figure 119).

Where clipper saws are of a 'wet' type, the dust is controlled by the moisture discharged. If a 'dry' type is used, a suitable extractor system is needed.

Protection, examination and storage of abrasive wheels

Abrasive wheels must be examined before each use, and especially after transit.

Forklift trucks

The following should be considered (and see pages 125-33):

1 The ground must be level and firm to avoid the forklift truck overbalancing

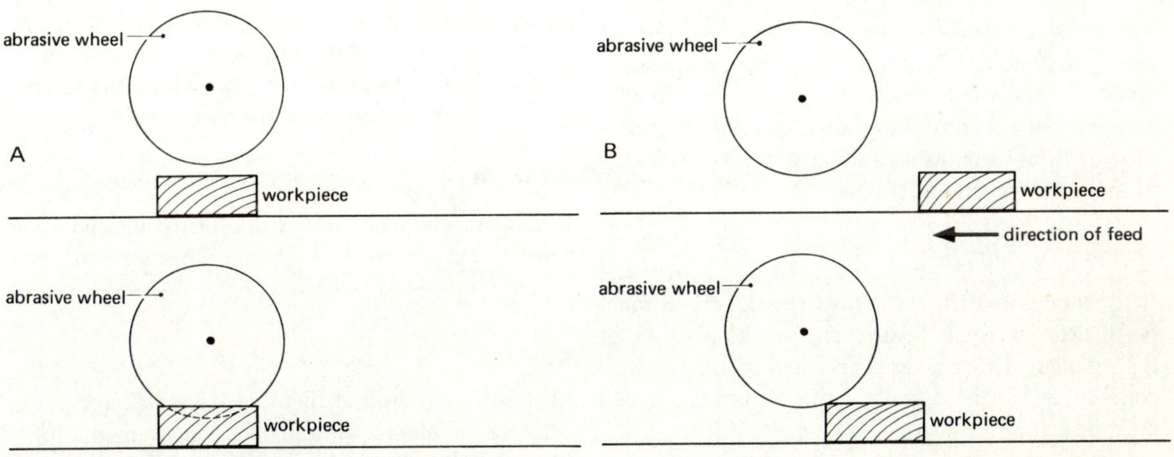

Figure 119 *Feed the workpiece to a clipper saw correctly.*
A – Right. The cutter is poised above the workpiece before the first cutting action. The cutter is lowered to commence cutting in the middle of the workpiece
B – Wrong. The cutter will hit the workpiece at a sharp corner, damaging the abrasive edge and causing pieces to fly off dangerously

2 All loads to the truck must be bundled and/or secured. Odd bricks or blocks should not be loose and easily dislodged
3 Scaffolds should be erected with off-loading from the forklifts in mind and additional tubing included
4 Drivers must be at least eighteen years old and competent to drive the forklift
5 The safe working load (SWL) must be clearly marked on the machine and observed

Legislation

Legal controls in the bricklaying trades mainly concern the safe use and adjustment of abrasive cutters.

The Abrasive Wheels Regulations 1970

Interpretation

'Abrasive wheels' mean cones, cylinders, discs or wheels used for cutting various materials. 'Overhang' means the part of the mandrel which is exposed between the collet and the nearest abrasive material.

Speeds of wheels

All abrasive wheels exceeding 0.055 m must be clearly marked to indicate the maximum permissible speed as specified by its manufacturer. Smaller abrasive wheels must have their maximum speeds shown on a clear notice, along with the overhang of mounted wheels and points.

Speeds of spindles

In conjunction with the above speeds, there must be a clear notice indicating the working speed of the spindle. This makes wheel and spindle speeds compatible. If the spindle is air driven, its speed must be suitably controlled by a governor.

Mounting of wheels

Only a trained, competent person shall mount any wheel, which must only be fixed to the machine for which it is intended. After the wheel has been brushed clean, inspected and accepted, the wheel is fitted, without unnecessary movement, and padded by washers of paper or similar compressible material. Where wheels with flanges are used, the flange must not be less than one-third of the total diameter. Clamping nuts must not be overtightened. Guards, when fitted must be secure and correctly adjusted to be as close as possible to the wheel.

Training and appointment of wheel mounters

Any person who mounts abrasive wheels must be trained and competent, and his appointment must be recorded by the occupier of the premises. A typical form, recommended by the Health and Safety Executive, is shown in Figure 120.

Guards

Their functions are to firstly prevent the operator from touching the wheel and secondly to contain any fragments should the wheel break up. The guard must enclose the wheel as far as possible. Adjustments to the guard must be made as the wheels reduce in size by wearing down.

Working rests

Where the work requires a supporting rest, it must be as close as practicable to the wheel, properly secured and substantial enough for the job at hand. Failure to adjust working rests can produce a serious hazard in jammed fingers and/or workpiece.

Advice

Within every workroom or workplace, where abrasive wheels will be used, a standard notice outlining dangers and precautions must be clearly displayed

Floors

All floor areas around the abrasive wheel apparatus must be maintained in a good, clean even condition with no rubbish or debris that may cause injury.

Employees' obligations

All persons using an abrasive wheel must do so in a safe and proper manner without wilful misuse or

Appointment of persons to mount abrasive wheels (Regulation 9)
Suitable form of Certificate

FACTORIES ACT 1961
THE ABRASIVE WHEELS REGULATIONS 1970
(SI 1970 No. 535)

Certificate of appointment

Name ____________________
is hereby appointed to mount the following classes or descriptions of abrasive wheels*:

____
____
____

SPECIMEN

Signature of occupier or his agent ____________________
Date ____________________
The above appointment is hereby revoked.
Signature of occupier or his agent ____________________
Date ____________________

*If all wheels are specified state "all".

APPENDIX 2

Appointment of persons to mount abrasive wheels (Regulation 9)
Specimen sheet for a Register

APPOINTMENT				REVOCATION	
Name of person appointed (1)	Class or description of abrasive wheels for which appointment is made (2)	Date of appointment (3)	Signature of occupier or his agent (4)	Date of revocation of appointment (5)	Signature of occupier or his agent (6)

Figure 120 *Specimen form issued to abrasive wheel mounters and register of certificate holders*

damage. All guards and protective apparatus must be used and personal protection, particularly goggles, worn.

General factors

The bricklayer should be aware of general safety procedures, particularly in relation to his fellow workers. Some hazards common to brickwork operations are as follows.

Cement limes

The powder content of these basic building materials are dangerous to skin and could lead to dermatitis and other skin diseases. Protective gloves must be worn and hand tools or other appliances must be used to handle these materials.

Mould oils

These release agents, used for concrete work, contain vegetable oils and will cause skin problems. If brush application should be undertaken, gloves must be worn at all times. Good personal hygiene is particularly important, before food breaks or at the end of each shift.

Bricks stacked on scaffolds

It cannot be overemphasized how careful the loading of a scaffold should be. Special attention is necessary to avoid excessive point loading or bad stacking of bricks and blocks that may collapse. High piles of bricks should be avoided and protective screens are necessary to all 'brickwork scaffolds'. (Typical details are described on pages 102-17.)

Cutting bricks on scaffolds

It is accepted practice to cut bricks on the scaffold platform. It is important, however, that all bricklayers watch where discarded pieces fall. There may be colleagues working below.

23 Painting and decorating safety

This chapter does not detail all the safety procedures of this trade. Other chapters have already covered many aspects applicable to the painter and decorator – scaffolds (pages 102-17), ladders (pages 118-24), etc. This section will highlight safety requirements particular to this trade.

Mechanical equipment

Rotary wire brush

This electrical powered hand-tool is used to clean off flaking paint or loose covering from metal surfaces, and in certain cases for de-scaling or de-rusting. When using it:

1 Always allow the moving head to stop before laying down the appliance to avoid reckless spinning
2 Always wear goggles or eye protectors
3 Where the size of any abrasive wheel exceeds 0.055 m the manufacturer must mark the maximum specified speed upon the appliance
4 All cables and connections must comply with minimum electrical standards (pages 134-41)
5 Retain a firm grip during use to avoid personal injury if the brush wanders or jerks
6 Mouth masks must be worn if dust is created

De-scaling pistol gun

This compressed air tool houses several needles which are forced forward on to a surface. The needles work as pistons, hammering repeatedly on to the surface for de-rusting, particularly on awkward areas of work. The following safety rules should be observed:

1 Always wear goggles or eye protectors
2 Do not carry out this work in an atmosphere that may be explosive as the impact of the needle on metal causes sparks
3 All hoses, couplings and connections must comply with the details on pages 146-8
4 Where rust or similar dusts are created, mouth masks or respirators must be worn

Rotary disc sander

Either electricity or compressed air drives a circular sanding head which supports abrasive paper, lambswool pad or in some types a carborundum grinding wheel. The uses vary according to the disc cover. The following safe practices should be observed:

1 Always wear goggles or eye protectors
2 Mouth masks or respirators must always be worn in dusty atmospheres
3 Connections, cables, hoses, etc. must comply with electrical or compressed air requirements respectively
4 A firm hold must be retained during use to avoid reckless or jerking movements which can cause injury
5 All pad facings must be secured with the correct threaded stud and pronged spanner

Burning equipment

Paraffin blow lamp

This consists of a fuel container on to which a pump and burner are attached to burn vaporized paraffin. This appliance heats and melts existing films of paint to allow them to be scraped off with hand scrapers. The following safety precautions must be observed:

1 During preparation and lighting, extreme care with relation to naked flames and flammable liquids must be observed
2 Body protection, particularly goggles and gloves, must be worn according to the position of the work and the intensity of the hazard
3 Always stand the blow lamp carefully, if a rest period is taken, and avoid any possibility of its overturning
4 Do not allow the work surface to become overheated, or allow fire to develop
5 Never leave a surface smouldering or in any danger of igniting and dispose of all burnt paint carefully
6 Always allow for any wind or draughts blowing the flame back at the operator

Gas torch

The torch is fed by liquefied petroleum gas (LPG) from a storage bottle, disposal cartridge or inside a self-contained torch. Irrespective of the type of appliance follow these requriements:

1 All LPG connections and hoses must comply with gas requirements (pages 201-4)
2 Always wear goggles and gloves as necessary for protection
3 Store all cylinders in a separate store, clear of all other hutting or storage with clear warning notices of contents – 'Keep clear: LPG'
4 Always turn off the cylinder control first to ensure the hose is fully drained
5 Never rest the burning torch on the cylinder or point it near the cylinder or any other fire hazard
6 Always allow for any wind or draughts blowing back the flame at the operator

Spray equipment

Shot blasting

This removes rust or millscale from metal or cleans stone or concrete. The appliance, powered by compressed air, blasts shot or grit with great force against the surface, the impact cleaning off unwanted particles. The following safety requirements apply:

1 All safety factors concerning compressed air must be observed (pages 146-8)
2 The hose must be reinforced rubber, with carbon black content to resist static electricity. Couplings must be of external type and retain a good passage of shot or grit
3 Always wear safety protectors to cover all danger zones. Goggles are inadequate, and a visor or face protector must be used with strong gloves, boots and a rubber or leather apron
4 Never aim the gun nozzle in any way that is reckless or dangerous. Aim only at the work surface
5 Always work in conjunction with manufacturers' instructions, regarding loading, using, pressures, etc.
6 Keep other personnel clear of the work areas by clear warning notices and barriers

Paint spray equipment

This evenly distributes paint to large areas, giving a smooth finish. The following must be observed:

1 All safety factors concerning compressed air must be observed (pages 146-8)
2 The hoses must be made of synthetic or natural rubber
3 Always keep the equipment clean and take special precautions when using solvents or other cleaning agents
4 Do not kink or bend the hoses
5 Always point the spray gun at the work surface and never fool with it
6 Always wear protective clothing, eye protectors and a good respirator. Mouth masks or similar cheap filters are not adequate
7 Where it is not practical to protect face and hands, always use a protective barrier cream before working
8 At the end of each working shift, or at meal breaks, make sure adequate body cleansing is carried out
9 Keep other personnel clear of any danger zone by erecting barriers and warning notices

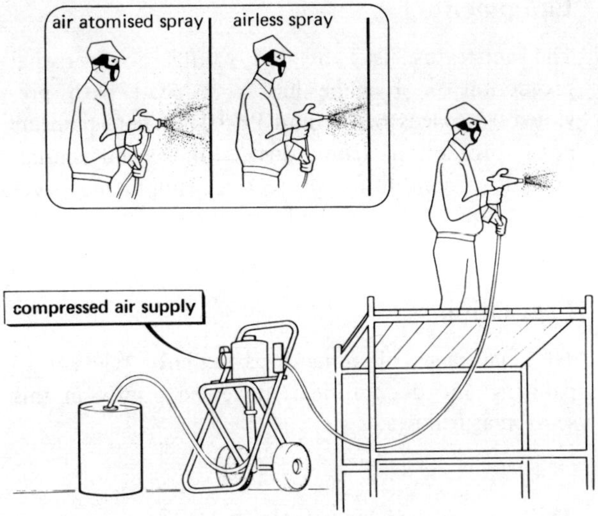

Note: Scaffold detail omitted for clarity.

Figure 121 *Airless spraying. It is called airless because the pressure of compressed air is used not to atomize the paint but to force it from a high-pressure pump through a narrow orifice which makes it atomize. The paint particles reach the surface with less force and bounce-back of paint is reduced (see inset). Waste of paint is reduced to a minimum. Application is faster. The system operates with long hoses, reducing the need to carry the apparatus up scaffolds and over hazards. The quantity of compressed air needed is reduced, together with the consequent hazards. No compressor is required*

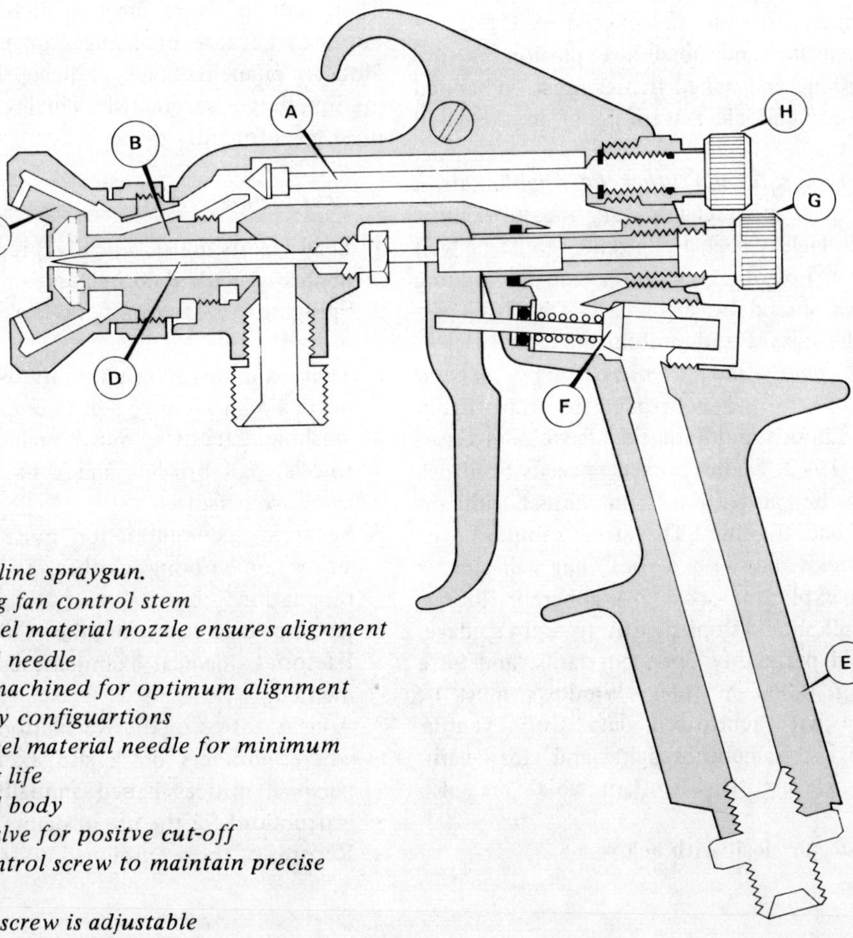

Figure 122 *Airline spraygun.*
A – Self-centring fan control stem
B – Stainless steel material nozzle ensures alignment of air nozzle and needle
C – Air nozzle machined for optimum alignment to maintain spray configuartions
D – Stainless steel material needle for minimum friction and long life
E – Drop-forged body
F – Nylon air valve for positve cut-off
G – Material control screw to maintain precise flow rates
H – Air control screw is adjustable

Materials

The following construction materials are specifically relevant to the safety of painting and decorating employees.

Paints and ancillary liquids

Oil-based paints, polyurethanes, varnish, etc. constitute no serious physical hazards, except in their storage. There is a possibility that body damage may occur if these materials are excessively inhaled. They must be carefully stacked to avoid injury at lifting or removing from store. A temperature of 15°C should be maintained, and stock used in order of its age to prevent old, unsuitable containers being stored and to avoid danger from corroded containers.

Spirit paint removers can be stored on shelving, but the temperature must not exceed 15°C because of heat expansion and possible explosion of containers. Smoking and naked flames must be banned in the vicinity and clear warning notices erected to this effect.

Paraffin, creosote and other flammables are a serious fire risk and precautionary measures must be observed. Tightly sealed containers must be well clear (6.096 m) of other store areas, or the building area. Notices should be erected to advise all personnel of the hazard and prohibiting naked flame.

Cellulose paint and thinners – compressed gases, petrol, etc. are controlled by the Highly Flammable Liquids and Liquified Petroleum Gases Regulations 1972, where storage exceeds 50 litres. There must be protection from direct sunlight and severe bad weather. The store should be of brick or blockwork, with a roof that will shatter easily if an explosion occurs. A concrete floor is required which should slope slightly to drain spillage. Doors should preferably open outwards, and be a minimum of 0.051 m thick. Windows must be strengthened with reinforced glass. Good ventilation, two exits, flameproof lights and clear warning notices stating 'Highly flammable' are also necessary.

Lead paints are dealt with below.

Equipment

The apparatus used by the painter for personal protection or hygiene has been dealt with previously – pages 63-71 and 96-101. Where painters need protection from dust, fumes and similar unhealthy conditions, protective equipment (pages 63, 68, 98, 100) is particularly relevant.

Legislation

The main legal document particularly relevant to painting and decorating is reviewed briefly in this section as follows:

The Lead Paint Regulations 1927

Their date of issue suggests these Regulations are outdated because of changes in modern processes. Modern pigments have reduced the lead contamination, but a severe risk remains and precautions must be enforced.

Duties of employers

1 Lead paints must be fully prepared and stored in clearly marked containers
2 Spraying of interiors with lead paint is not permitted
3 There are restrictions about using abrasives on lead paints to reduce risk to employees
4 Washing facilities must include water, soap, towels, nail brushes and a basin or bucket for each five persons
5 Separate accommodation must be provided to store non-working clothes, to avoid lead contamination
6 At the discretion of the Chief Inspector of Factories, medical examinations must be implemented
7 A clear notice of the Regulations must be issued where numbers on a site exceed twelve, and personal notices issued annually on the health instructions for the use of paints
8 Records of those employed must be kept

Duties of employees

1 Overalls must always be worn, washed at least weekly, and not worn during food breaks
2 Strict observance of protective measures is required from all employees
3 All employees must safeguard their clothing from contamination by using the separate storage facilities provided
4 Good hygiene must be observed, particularly when using lead paint, and strictly enforced at food breaks or on leaving work
5 Any employee must attend any prescribed medical examination when requested to do so

General factors

1 Paints used on objects which may be sucked or chewed by children must not contain more than 1.5 per cent of lead or similar known toxic compound (details in BS 4310: Permissible Limit of Lead in Low Lead Paints or Similar Materials)
2 Those applying paints to any surface must always observe the requirements of that surface area. Degrading or lowering of a functional requirement must not happen, e.g. adjustment of the spread-of-flame factors
4 Adhesives or paints containing petroleum spirit must be used with caution to reduce fumes
5 When pastes or adhesives containing fungicides are used extreme caution is required, especially in personal hygiene. It is difficult to protect hands from this work but keeping them clean constitutes reasonable care

24 Mechanical services safety

Services installation is carried out by plumbers and heating engineers, with Gas Board service engineers establishing a supply of gas. These two groups of employees are more inter-related to construction work than telephone and electricity installation groups.

Mechanical equipment

Electrical drill

This commonly used appliance has been considered (pages 134-41). When used as a fixed machine on a stand, usually in workshop conditions, the following basic safety factors apply:

Entanglement: the chances of the operator becoming entangled within a drill are influenced by his personal dress. Everyone engaged in drilling work must keep hair well clear of the moving parts, either by wearing a cap, or by having a shorter hair style. The wearing of jewellery must be discouraged and where bandages indicate any form of danger, a transfer from drilling work must be considered. Any loose clothing must be kept well clear of the machine, particularly shirt sleeves

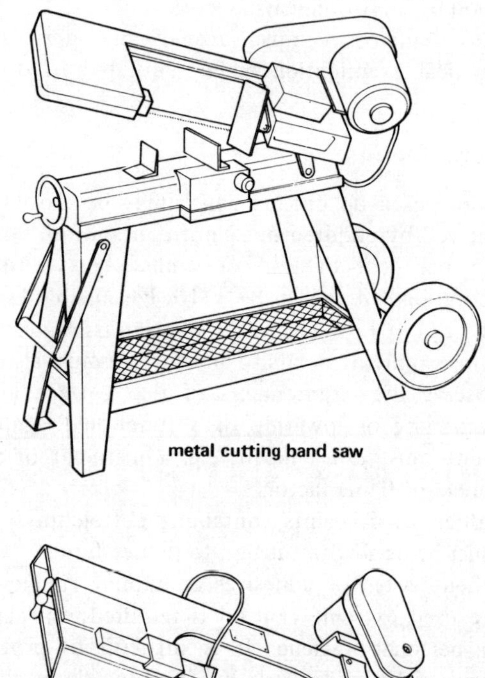

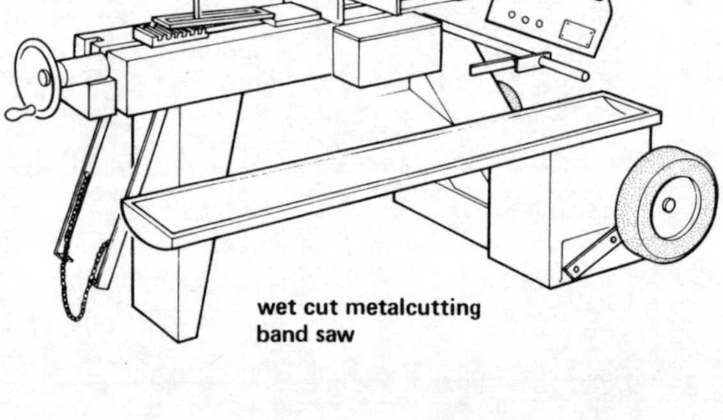

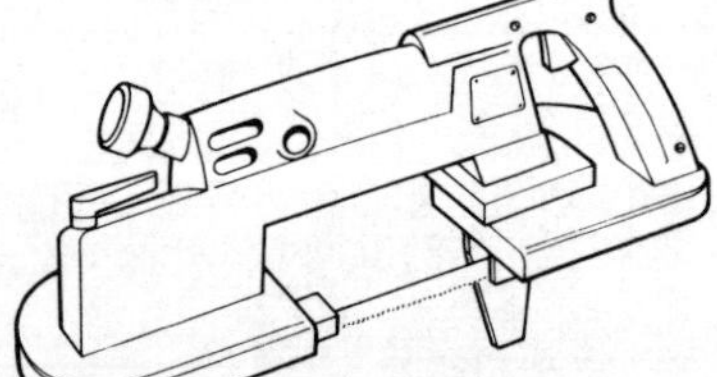

Figure 123 *Electrical cutting appliances*

which must be rolled up tidily. Gloves must not be worn.

There are numerous ways of guarding the moving parts; an effective protector must be fixed to all drilling machines. The method used matters little provided the drill is completely covered, particularly at its resting position at change of a workpiece. The operator's view of his work must not be restricted and all guards must be adjustable to achieve the best position for any different workpiece.

Clamping: all work must be clamped in position prior to drilling. Failure to do this results in the spinning or falling of the workpiece, which can injure the operator and/or anyone else in the vicinity.

Checklist

1. Always use the guards provided
2. Check the machine for stability prior to use
3. Clamp all work before commencing
4. Keep hair short or protected from entanglement
5. Do not clear away swarf with bare hands
6. Keep clothes and hair tidy and well clear of the drill
7. Make sure all belts or moving parts are guarded, in addition to the cutting parts of the machine
8. Remember to remove the chuck key after using it
9. Never wear gloves for drilling
10. Use lubricants if needed to improve cutting

Guillotines

The most common form of guillotine used by services personnel is to cut thin gauge metals. A typical example is shown in Figures 126 and 127 and the following points need consideration:

Front guards: all working must be carried out from the front of the guillotine. There must be an effective guard to prevent fingers from getting under the blade. This is particularly dangerous where two people operate the machine and a misunderstanding may occur. Where the workpiece is

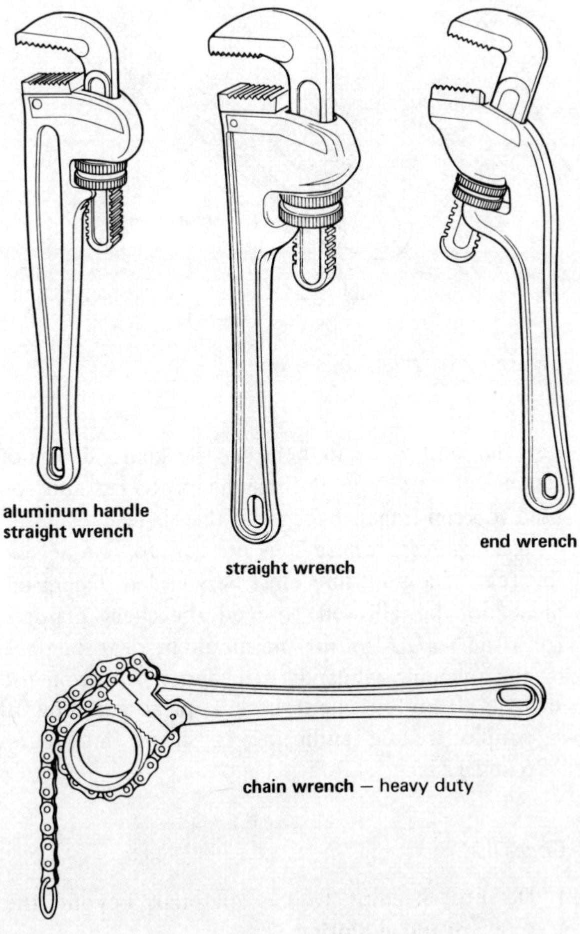

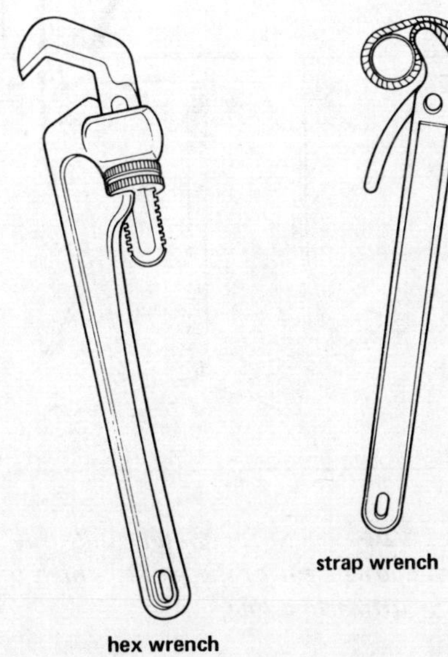

Figure 124 *Hand-operated wrenches*

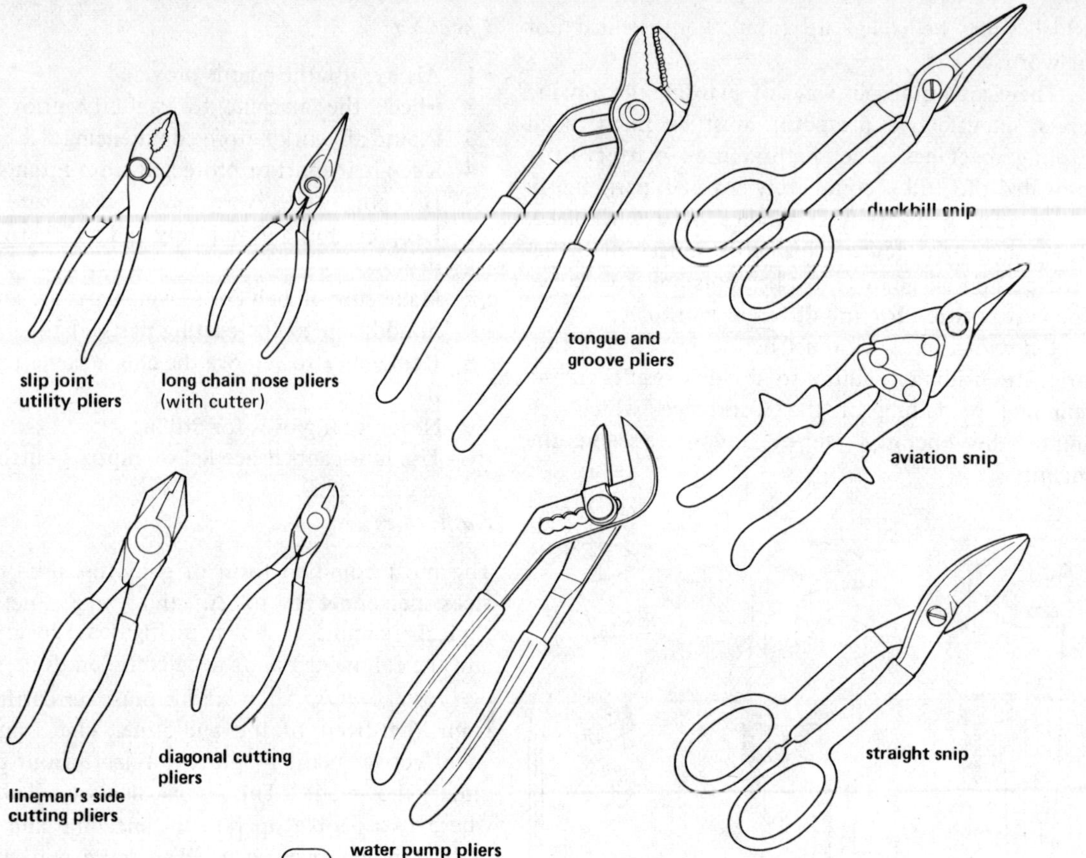

Figure 125 *Pliers and snips*

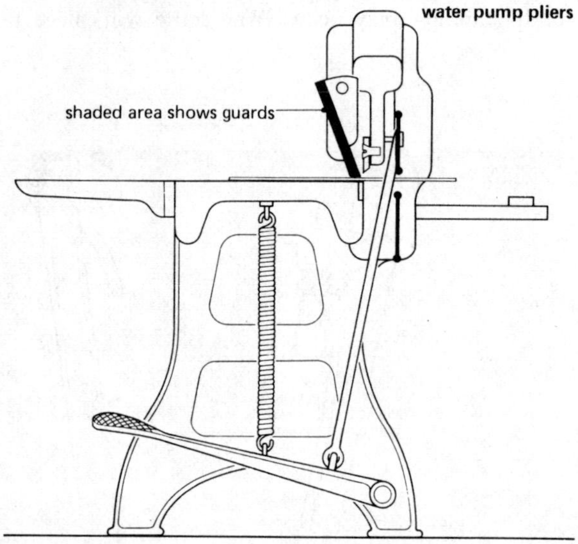

Figure 126 *Sheet metal guillotine – the front guard allows sight of the work, which is needed when cutting to a line*

not flat and the gap beneath the guard does not restrain the passage of finger, a protector must be used to keep fingers back from the blade.

Rear guards: where it is possible to gain access, the rear of a guillotine must be guarded. Operators must not be allowed to feed the sheet material from the rear. Also, no one should be near the back of the machine whilst it is in action, to avoid the dangers from cut materials. Typical guarding of a simple treadle guillotine is shown in Figures 126 and 127.

Checklist:

1 Do not attempt to cut materials beyond the scope of the guillotine
2 Keep the machine clear of waste

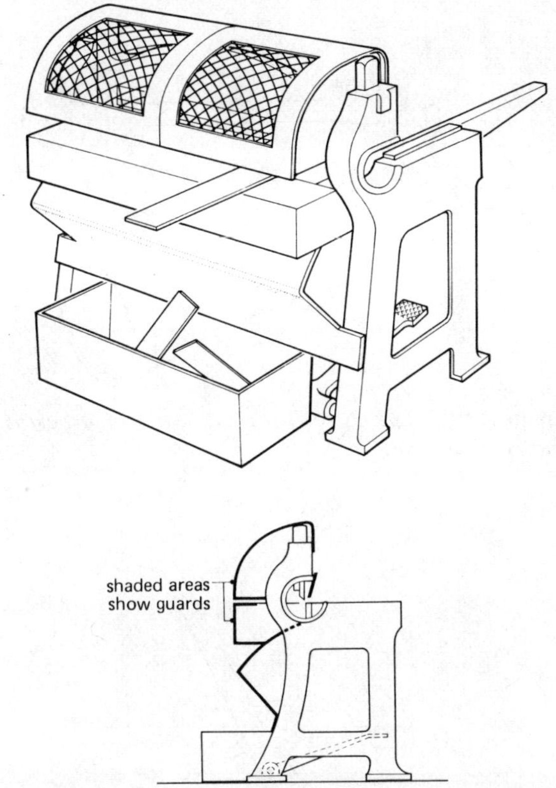

Figure 127 *Sheet metal guillotine – rear guard. The guard provides a feed opening for stock and a lower opening, with deflector plate, for the cut pieces which would be pushed off the table beneath the front guard*

3 Never attempt to use the machine without adequate guarding
4 Never adjust the machine whilst it is connected to the power source
5 Do not rest your feet on the treadle between cuts, in case of misuse or accidents
6 Always wear gloves to avoid minor hand injuries
7 When first using the machine, test it to get the feel of the treadle and its spring tension

Threader

This machine is often a portable type used for cutting threads joining pipework. The following should be observed:

1 Keep all moving parts guarded
2 Where long lengths pass through the machine,

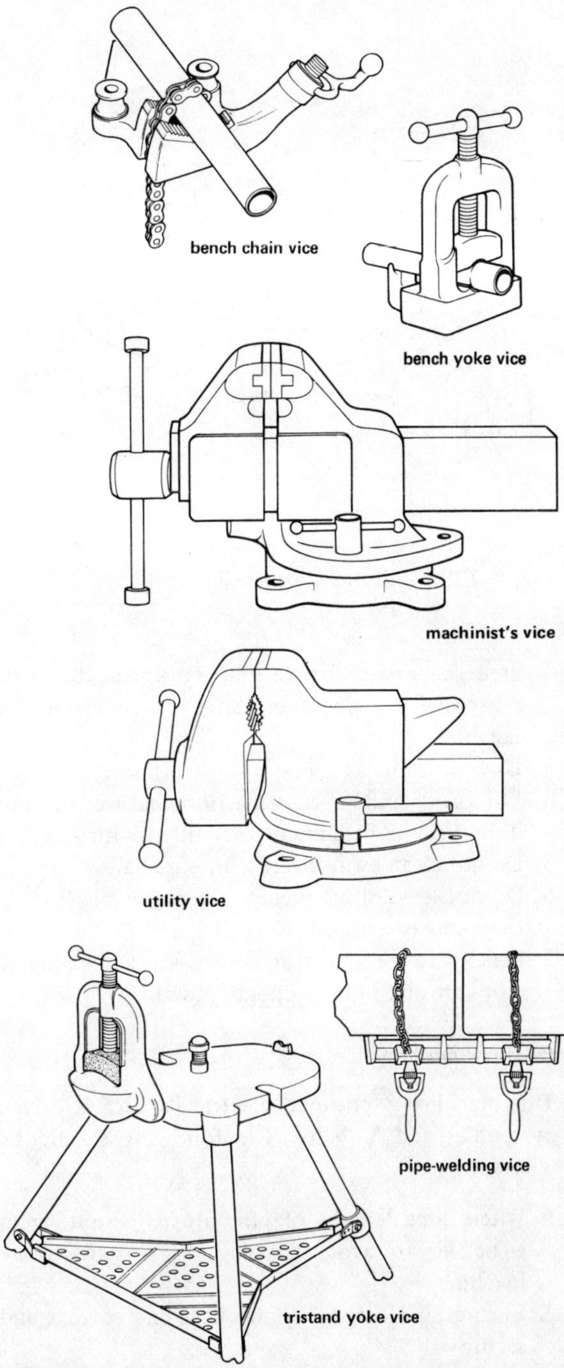

Figure 128 *Service engineer's vices*

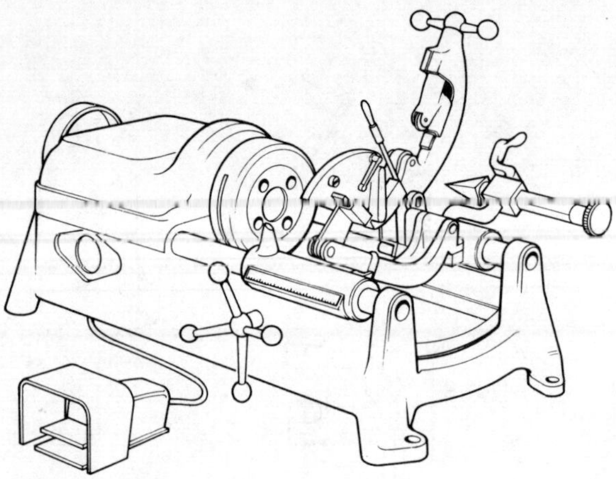

Figure 129 *Automatic threading machine*

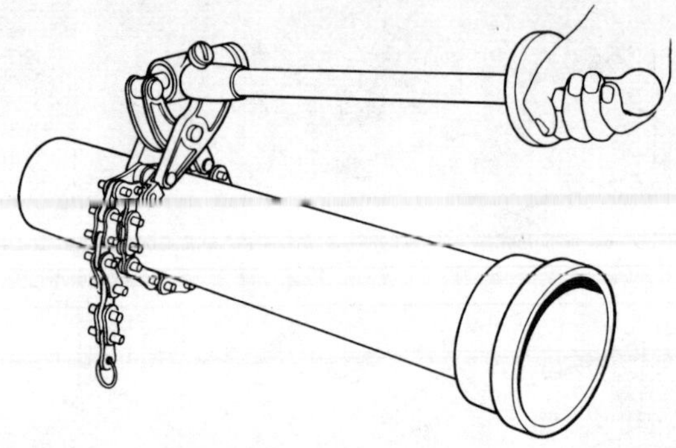

Figure 130 *Pipe-cutter for clay, cast iron, asbestos and cement pipes*

erect a barrier around the workpiece to avoid personnel injuring themselves or damaging the machine

3 Use lubricants as necessary to work safely
4 Lubricant spillage causes a floor hazard; prevent it by using a catchment tray filled with sawdust
5 Do not clean away swarf with bare hands
6 Do not leave short pieces of pipe on the floor – they may be tripped over
7 Make sure the operator knows where emergency stops are and how to operate them

Powered hacksaw

This machine is powered electrically and is driven by pullies and V belts. The following should be observed:

1 Where long lengths of pipe are to be cut, erect a barrier to avoid tripping or damage to the machine
2 Ensure all electrical connections are correct and secure
3 Make sure all moving parts are guarded. Where the belt is exposed, it must be guarded (although this is not so vital where the belt is moving outwards to push away the fingers)

Figure 131 *Assembly tool for gasketed soil pipe*

4 Always use sharp blades, free from defects
5 Do not allow the workpiece to fall a long way at the completion of cutting. Use supports
6 Make sure emergency stops are known and function

Materials

Several of the materials used by service installers can be a hazard if not used carefully.

Lead

Lead poisoning is the danger to be avoided. Page 68 explained poisoning both through the mouth and through skin pores. The following should be observed:

1 Wear protective clothes and remove them at food breaks or at the end of each day
2 Use barrier creams on hands
3 Always wash your hands, using scrubbing brushes, before any food break and at the end of each day
4 Avoid smoking because lead-contaminated air is inhaled at this time
5 Do not leave off-cuts of lead. They may be dangerous to children, particularly on domestic installations

Asbestos

The full range of protection (pages 64-6) should be observed to reduce health dangers.

Compounds

Various compounds are used in jointing. Skin diseases may follow if these simple basic precautions are not observed:

1 Never apply compounds with bare hands. Use small wooden spreaders
2 Wear protective clothing, and remove this at food breaks and at the end of each day
3 Always scrub your hands before food breaks and at the end of each day
4 Keep compounds away from children, particularly on domestic installations

Equipment

Non-mechanical equipment is mainly used for pipe-cutting, bending or threading

Pipe vice

This is usually a bench-mounted vice to secure pipework for cutting or for hand-threading. The vice is relatively harmless but the following apply:

1 Make sure the vice is secured properly to the bench
2 Set the vice correctly for the workpiece
3 Do not exert unnecessary pressure on the vice
4 Never improvise or use the vice as a bender
5 Where the vice is free-standing, ensure a firm base area is used

Pipe bender

This portable tool is a lever for bending pipework to predetermined radii for installation. The following should be observed:

1 Make sure the equipment is erected on a firm base
2 Use only the correct radius former for bending
3 Take a good stance for levering the pipework to avoid personal injury or body damage
4 Allow adequate room around the equipment for the length of pipe

Hand threading

Where it is not economical to have powered threading appliances, the hand-operated die cutter is often used on site. The following apply:

1 Allow the cutter to act freely, and take out the thread in easy stages
2 Make sure the workpiece is secured
3 Do not remove swarf with bare hands
4 Use lubricants as necessary to help cutting, but never apply this with bare hands
5 Provide a barrier around long lengths of pipe to avoid other persons being injured or the equipment being damaged
6 Always clean down equipment after use

Liquefied Petroleum Gas (LPG) and appliances

This is among the most frequently used and most hazardous equipment in services installation. The following review is intended to explain all safety requirements applicable to gas and the appliances involved with the use of gas:

LPG

The full title is Liquefied Petroleum Gas, and it is also known commercially as butane or propane. It is stored under pressure in liquid form and supplied in containers specially designed to withstand vapour pressures. The colour of each cylinder varies according to the supplying company.

Connection to the appliance is achieved by a male or female connector with plastic protectors to each threaded part. These caps give an additional safeguard should the valve be accidently turned on. A wide range of cylinder sizes are available according to the requirements of different workplaces. All cylinders must be transported upright with a sling or similar device to prevent movement or bumping. Because the gases are heavier than air, they can disperse at ground level or even to underground pipeworks or basements. The storage and handling of LPG must observe the following:

1 The store must be easily accessible and clear of underground areas
2 Storage must be on a firm level base
3 Exposure to extreme hot or cold weather must be avoided
4 Combustible materials must be stored well away from the gases

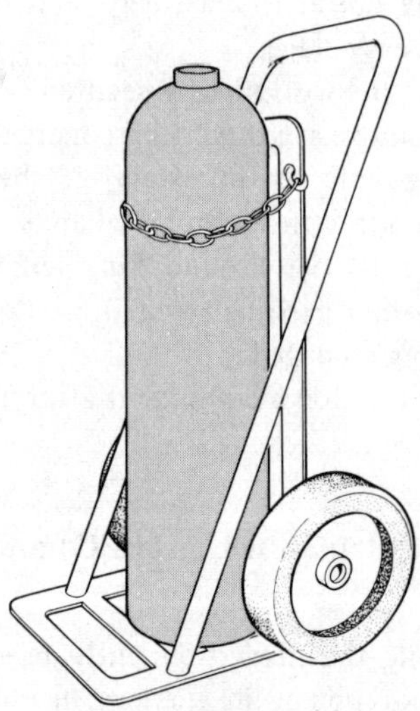

Figure 132 *Cylinders are secured to the trolley by chains and cannot roll about or fall off*

5 Full and empty cylinders must be segregated with valves shut and protective caps fitted
6 Cylinders must be kept, with the valve to the top, in a secure store area, with clear notices to prohibit smoking – 'No Smoking and No Naked Flame'
7 Cylinders, if too heavy to lift, must be handled on trolleys or skids
8 Do not lever or pull on valves during handling or storage
9 A good supply of dry powder fire extinguishers must be available
10 All stock of cylinders must be controlled and recorded by a good stores procedure. Damaged or defective cylinders, and those involved in a dangerous situation, must be returned to the supplier. A typical notification form is shown in Figure 133

LPG cylinder connections

To achieve a safe working procedure the following activities must be completed carefully:

1 Stand the cylinder on a firm base clear of ducting, trenches or basements
2 After checking that the valve is shut remove the protective cap and keep for future use
3 Ensure that the cylinder and regulator threads are not damaged
4 Fit the regulator to the cylinder, making sure the hose is slack to avoid any force or cross-threading. Fit it by hand initially and tighten, without unnecessary force, with a spanner of correct size. (The thread, being left handed, turns anticlockwise and is indicated by a groove to the cutter corners of the securing nut or coupler)
5 Do not use any form of sealing compound for cylinder connections
6 Never use oil or grease on threads of gauges or cylinders
7 Test the connection, as described below, before commencing work

LPG – inspections and testing

Inspections must be carried out at regular intervals. If any leak is detected which cannot be controlled, the offending cylinder must be taken to an open space, well clear of buildings, ducting, underground pipework or drains, etc. With the leak uppermost, the cylinder must be marked 'faulty' or 'dangerous – leaking' and the supplier contacted immediately. The following is the procedure for testing and inspection:

1 Check for leak by smell or listening. Also check by touch: a leaking cylinder is cooler than a safe cylinder
2 If a leak is suspected, cover all joints and connections with soapy water. If bubbles appear, there is a leak
3 If a leak is apparent, turn off the valve and reconnect the defective joint
4 Recharge the appliance and retest. If the leak is still apparent, change the fittings or cylinder until you have a good leakproof appliance

R & D Building Company

Pararad Road, Roselip

TO ....

The following details are referred to you as confirmation of a recent fault/accident/dangerous occurence involving storage cylinders.

Date of incident	Serial No. of cylinder	Contents of cylinder	Owner's identification	Comments

Figure 133 *Form for reporting mishaps with gas cylinders*

LPG – ignition of gas for use

1 Have matches or lighter ready for ignition
2 Ensure all valves are shut
3 Set the regulator to the correct pressure
4 Open the valve and ignite the escaping gas
5 If the first attempt fails or the igniting match goes out, *allow the escaped gas to disperse* before a second attempt. DO NOT ALLOW GAS TO CONTINUE TO ESCAPE BETWEEN ATTEMPTS – *an explosion may occur*
6 If any part of the appliance fails during ignition, turn off the main valve and check all parts of the equipment

LPG – the working process

The operator should keep a close watch on all parts of the appliance and observe the following checklist:

1 Keep adequate ventilation for cylinders, burners and personal health requirements
2 Avoid excessive draughts that may extinguish the flame
3 Maintain a check of all controls, for ease of movement
4 Keep the regulator set correctly either for ignition or operating
5 Maintain and check all equipment with particular precaution against damage during use
6 Keep all combustible materials clear of the working area, at least 0.920 m away. If this cannot be achieved, protective materials must be used

LPG – turning off after use

Always turn off the *main valve first* to allow the gases within the supply hose to be burnt. Do not turn off valves at the appliance, as this leaves the hoses charged with gas.

Disconnecting from cylinders

After completion of work or when the cylinder becomes empty, the following should be observed:

1 Turn off all valves
2 Remove regulator and other equipment and store carefully
3 Refit the protective cap
4 Return the cylinder to its store with a cap sealer securely fixed. Ensure the empty cylinder is returned to the correct bay. Full and empty cylinders must remain segregated

Precautions at flame failure

Causes of flame failure, and their cures, are as follows:

1 *Failure of equipment:* shut off main valve
2 *Too much gas:* turn off the main valve and re-ignite, as previously described
3 An *empty cylinder*: turn off main valve and replace cylinder
4 Too much gas caused by *incorrect regulator setting:* adjust regulator
5 *Excessive draught* forces out the flame: prevent draughts before re-ignition
6 *Dirt* within the appliance or hose: shut off main valve. Dismantle and clean out appliance and hoses as necessary

General factors

Hygiene

The main problem encountered by services trades is keeping clean after the use of lead, compounds and similar materials that cause personal dirt or poisoning problems. The need for barrier creams and good personal protection cannot be overemphasized. Hand scrubbing brushes must be provided and all employees should use these to safeguard against infection.

Oxygen

It is extremely dangerous to breathe oxygen directly from a cylinder. This must never be done.

Protective clothing

Pages 207-9 give full information about body protection during welding and cutting.

25 Welding safety

Oxy-acetylene welding

In this process a flame is produced by a balanced supply of fuel gas (acetylene) being in a supply of oxygen to create combustion. This flame is used to melt the metals involved with welding.

Materials

Fuel gas and oxygen (or compressed air) is stored and handled in strong metal cylinders referred to just as 'cylinders', but correctly termed 'compressed gas cylinders'. The following hazards may exist:

1 *Mechanical damage* to the cylinder, causing leakage
2 *Heating*, causing an increase in internal pressure or a weakening of the cylinder
3 Other *substances in contact* with the gases setting up a dangerous chemical reaction, e.g. oil and oxygen or copper and acetylene
4 *Leakage* may occur and the gas ignite if confined to a poorly ventilated space

Manufacture of cylinders

Strict statutory powers control the materials, methods and design of cylinder manufacture. Corresponding care must be taken with the cylinders in use. Interchange of gases to cylinders should be avoided, and BS 349 recommends the standard colour code and identification details (Figure 134). In addition, combustible and non-combustible gases are kept separate by using different screw-handing of the valve outlets. The following summarizes the details of cylinders used in construction site welding.

Nature of the gas or mixture	Colour of bands: Nominal	Colour No. to BS381C	Container neck
non-flammable and non-poisonous	none	none	
non-flammable and poisonous	golden yellow	356	yellow
flammable and non-poisonous	signal red	537	red
flammable and poisonous	signal red and golden yellow	537 and 356	red / yellow

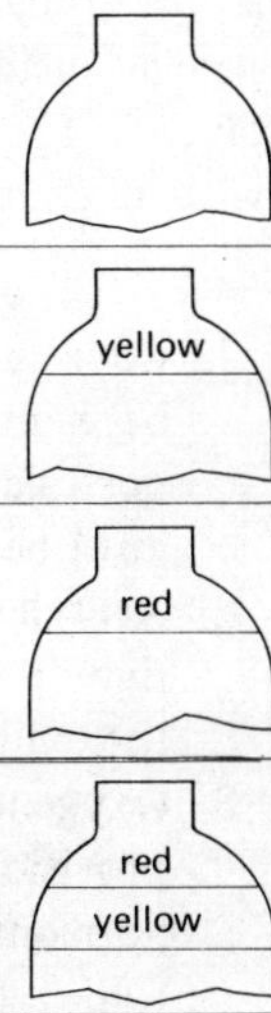

Figure 134 *Colour-coding of gas cylinders. Bands on the cylinder neck indicate hazard properties*

Gas	*thread of valve*	*colour of cylinder*	*characteristic*
compressed air	right hand	grey	non-combustible
oxygen	right hand	black	non-combustible
acetylene	left hand	maroon	combustible

Checklist

1 Never temper with or use valve threads that have been adjusted
2 Always check the labels on cylinders at delivery and before use
3 Refer to each gas by its proper name, e.g. never refer to oxygen as 'air'
4 Observe the requirements of BS 349 and statutory controls

Storage of cylinders

A systematic store procedure is important. Different kinds of cylinders must be kept separate, with a special distinction between full and empty containers. The following checklist must be observed with all compressed gas cylinders:

1 Combustible and non-combustible gases must be kept in separate rooms. Where this is not possible, a clear division between each type must be established
2 Acetylene must always be stored in the upright position with provision made to restrain the cylinder from falling
3 Oxygen can be stored upright with the same provision as for acetylene. If oxygen is stored horizontally, large wedges should prevent rolling (Figure 135). No more than four cylinders high is allowed, and large cylinders should be at the base of the pile
4 Full and empty cylinders must be kept in different sections of the store with clear notices indicating which each store section contains. Confusion must be eliminated
5 Storage areas must be well ventilated at top and base to permit a clear passage of fresh air
6 The cylinders must be kept away from extreme weather conditions, e.g. direct sunlight or severe frost. A tarpaulin is not adequate as protection
7 Avoid corrosion by keeping cylinders away from moisture and wet soil
8 All lighting in combustible gas stores must be flame proof or positioned outside the store room and providing light through a fixed window panel
9 Store well away from any source of heat, e.g. store radiators. The heat may cause weakening of the cylinder or increase in gas pressure and subsequent danger
10 Keep all cylinders, especially oxygen, well clear of any contamination of oil or grease
11 Keep all cylinder valves clean of dirt, grit, oil or similar contaminants
12 Always record any dangerous occurrence to the cylinder owner (see Figure 133, page 203)
13 Never smoke or use any naked flame near or within the store area
14 There must be clear notices indicating the storage area, its contents and warning against naked flame – 'LPG – No Smoking'
(LPG is the standard abbreviation for Liquefied Petroleum Gas)
15 Never roll cylinders or use them to aid movement of other equipment
16 Maintain a clear access and exit from the store area, should emergency conditions arise

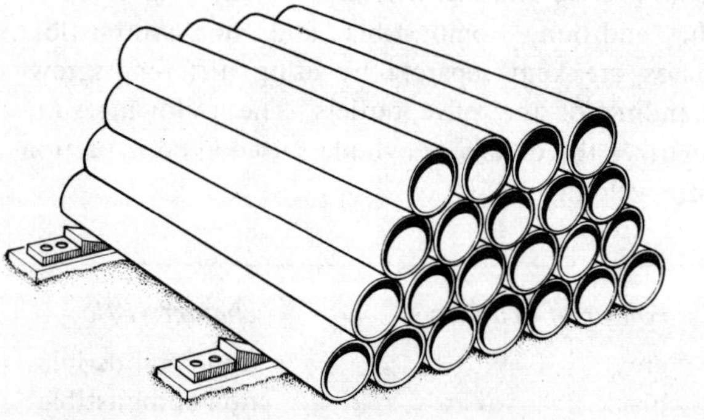

Figure 135 *Acceptable horizontal storage of gas cylinders*

Handling of cylinders

Handling occurs when cylinders are delivered or moved from the store to the work area. Many of the hazards listed above apply here and must be observed in addition to the following checklist:

1 Do not allow cylinders to be bumped or dropped so they may cause mechanical damage, i.e. not projecting behind or to the sides of any vehicle
2 Never transport cylinders loose. Slings eliminate bumping or dropping during transit

3 Always off-load individually with a sling, or by crane with a cradle of non-slip materials. Do not lift by any nozzle or valve appliance
4 Never handle or adjust oxygen cylinders, valves or fittings with grease-contaminated hands, gloves or rags
5 Keep cylinders away from electrical wires and appliances
6 Never allow cylinders near to spark, flame or welding operations that might cause heat or flame
7 Do not lubricate any valve, especially avoiding lead-based jointing compounds
8 Avoid dirt or grit entering the valve. If this occurs open and close the valve momentarily (snifting) before attaching fittings or regulators (see also 'Equipment' below)
9 Never transport cylinders whilst hoses and regulators are attached unless on a suitable trolley (Figure 132, page 202). Always keep the valve shut during transit
10 Keep clear of sources of corrosive materials, e.g. battery charging or similar hazardous areas

Equipment

Provided the storage and transport facilities have maintained all cylinders in good condition, the operator is required to establish a safe working procedure for himself. The following checklist must be adhered to:

1 Do not increase the leverage upon keys by employing large or long-handled spanners
2 Do not use broken or worn keys
3 Never attempt to get gas from broken spindles
4 Never attempt to fill one cylinder from another
5 Open the cylinder valve slowly and never over-tighten to shut off
6 Do not attempt to thaw a frozen valve with flame; use hot water
7 If deposits of grit or loose dirt are suspected within the vale always 'snift' before connecting fittings or regulators
8 Never attempt to work if automatic pressure regulators are not fitted to both gas cylinders
9 Before fitting a regulator to a full cylinder, release the adjusting screw to avoid damage to the regulator
10 Check that all connections are compatible before fitting them, and do not force the connections to fit
11 Always use good-quality hose, free from hardening, cracking or similar defects
12 All hose connections must be securely attached without any leak
13 Joints in hose should be avoided, as should unnecessarily long hose. All jointing must be achieved by proper connections
14 Always allow an adequate supply of fuel gas from the blowpipe nozzle before lighting up
15 Always check for leaks with caution and do not continue working if one is suspected
16 Where a leak is suspected, test with an application of soapy solution mixed with water, brushed on to all joints and connections. Bubbling of the solution indicates a fault. Never test for leaks with a naked flame

Personal protection

Much of the personal protection applicable to welding operations has been reviewed (pages 63-71, 96-101). The welder is vulnerable to both health hazards and physical body dangers.

Contamination of lungs

A serious problem caused by vapours, gases, fumes and dust arising from:

1 Surface cleansing and preparation for welding
2 Action of the heat source upon air
3 Action of the heat source upon base material, its castings or welding rods

Surfaces can be prepared with abrasive or chemical cleaning to remove scale, rust, dirt and general matter. A local exhaust mechanism best eliminates the pollution immediately it is created, i.e. at the face of the workpiece. There is an additional hazard of poison from sealed storage tanks (page 210).

The heat source may activate nitrous fumes that may be emitted into the air. These can be a problem in confined spaces. The application of heat to base materials, or their coatings, generates fumes including particles of respirable size. Most particles produced by welding metals are relatively harmless. However, coatings used to protect metals give off harmful fumes that should be eliminated before welding by good surface preparation. These dangerous fumes are produced by lead-based paints, zinc galvanized and similar metallic coatings or the presence of grease and/or oil.

To cure these problems there must be good ventilation. It is important, first, to realize that the size of particle to be inhaled has a strong influence upon its danger to the inhaler (Figure 137, page 211). Finer particles, which can pass through the natural body barriers, are the most hazardous. Ventilation can be achieved by:

1 General ventilation
2 Local ventilation
3 Confined space ventilation
4 Personal protection

General ventilation can be achieved in large areas where welding is done behind flash screens or behind flash curtains. Natural ventilation through windows or roof lights is not satisfactory: a mechanical extraction system must be used. All harmful products in the air are thus removed from the workplace and a clean working environment is retained.

Local ventilation is an exhaust system that removes the contamination at its source. An exhaust hood, no more than 0.500 m from the workpiece, is connected to a 0.125 m diameter flexible, reinforced, fire-resistant hose. This propels the contaminated air to an electrostatic precipitator which cleans the air. The unit can be wall-mounted or portable to meet the requirements of each workshop, and retains a maximum noise level of 70 dB, so no ear protection is needed whilst in operation.

Confined space ventilation is the term given to conditions that have poor or restricted natural ventilation. A build-up of harmful gases and fumes can cause a serious hazard and must be avoided by supplying fresh air. It is important that oxygen is not used as a means of air enrichment. Any increase in oxygen increases the fire hazard and must be avoided.

Personal protection is required where no other means of avoiding the fumes is possible or practical. A respirator worn by the welder filters the air before it causes personal danger. Pages 64-5 give details of suitable styles and types.

Body

To avoid hot metal particles spraying on to personal clothing or the body, a leather apron must be worn. This should be full length and secured to the body for a good close fit from chest down to ankles.

Hands

To protect the operator from spatter, heat and general welding dangers, a gauntlet style of leather glove must be worn. Where excessive heat is expected asbestos gloves may be necessary.

Feet/ankles

Below the apron level, the feet and ankles may not be adequately protected from molten metal. Spats of leather or asbestos cloth should be worn to avoid burns to the body at low level or damage to clothing and boots.

Head

To protect the head, a leather cap or helmet should be worn. Where excessive heat is anticipated a face shield should be used to protect both eyes and face completely.

Eyes

A welder must always wear eye protection. Goggles provide protection from flying particles and the intense flare given off by welding. Details of the welders' goggles are incorporated with BS 679. See also pages 97-9, 169-70).

Other people

To protect other persons near welding activities, screens or welding curtains must be used to enclose the work area. This should not, however, restrict ventilation or movement within the work area, nor the passage of fresh air into the work area.

Hazards and precautions

Several processes automatically follow the preparation activities. The following checklist is intended for all those who carry out welding operations:

1 There must be ample ventilation. Where the location of the workpiece prevents this, special precautions must be taken (pages 64, 208).
2 Working areas must be clear, particularly the floor
3 All combustible materials must be removed and combustible fixtures or fittings protected from welding flame or molten metal particles. A clear distance between workpiece and combustibles must be maintained
4 Do not weld on or near timber structures unless adequate protection has been applied
5 Check equipment, regulators, hoses and connections regularly
6 Use a colour code for hoses, e.g. blue or black for oxygen and other non-combustible gases and red hose for acetylene or other combustible gases
7 Keep hoses of similar length
8 Do not wrap surplus hose around cylinders
9 Never use copper or copper alloy in acetylene hosepipe for connections or fittings
10 Maintain a pressure within that recommended by the manufacturers in accordance with the regulators. Check the pressure regulator regularly
11 Always light up in the correct sequence: allow sufficient time for the hose to be clear of air and charged with acetylene before attempting to ignite
12 Keep hoses clear of heat, grease, oil or passing traffic
13 Where a leak is suspected, test by covering with soapy water. Never test for a leak with a naked flame
14 Never leave a blow torch hung on to or near the cylinder nor lying in the vicinity of empty or charged drums of any liquids

Flashback

A flashback is created by the flame reverting into the charged hose, causing explosion of mixed gases. It is usually associated with a disconnection of regulators or blowpipes or when new hoses are used for the first time. Loose connections are often blamed. The following actions may avoid flashbacks:

1 Ensure all connections are tight
2 Always close each gas valve separately
3 Charge each blowpipe separately by purging, allowing only one gas valve to be opened at any one time
4 Make sure regulators are set to correct pressures
5 Always be sure that blowpipe valves are closed and cylinder valves open

To correct the problem and avoid further dangers:

1 Close all valves to blowpipe and cylinders
2 Drain hose
3 Clean, adjust, repair and replace equipment as necessary

Flame snap out

Snap out is an accidental extinction of the flame outside the nozzle of torch. It can be caused by incorrect pressures of regulators or an inadequate flow of gases with the valves partially closed. The torch may also be obstructed if the welder's action in holding it too close to the work extinguishes the flame. Avoid these faults, but if they occur:

1 Shut off torch valves
2 Check the settings of regulators
3 Check pressures of cylinders
4 Check nozzles of torch, relight and maintain an adequate flow of gases

Heated cylinder

A dangerous situation is created where acetylene cylinder or cylinders have become heated or a serious backfire has occurred. The cause may be faulty equipment allowing a backfire or bad positioning of cylinders. The action required is quick and decisive. Follow this sequence:

1 Raise the alarm

2 Clear away the heat source and shut off valves
3 Remove all fittings and drag the cylinder to open space
4 Open the cylinder fully, and apply plenty of water

In the meantime all personnel should be cleared away and the fire brigade called. Subsequently the owners of the cylinder should be contacted and the cylinder removed (Figure 133).

Backfire or sustained backfire

These incidents can occur at the lighting up stage or during use. In both cases the flame is allowed to enter the hose where it is either self-extinguished (backfire) or maintains a flame (sustained). The faults usually arise from incorrect regulator setting or an attempt to light up before the gases are balanced. In certain cases the torch may cause a problem because the nozzle is obstructed or heated. To correct backfire:

1 Close off blowpipe valves, oxygen first
2 Check pressures of cylinders
3 Adjust and check regulator pressures
4 Cleanse, check and prepare the torch
5 Establish the flow of gases and relight

Working on painted surfaces

A hazard that creates serious dangers of breathing in harmful fumes or receiving damage to eyes from glare or impact of metal particles. The following should be observed:

1 Wear a respirator (pages 64-5)
2 Provide good ventilation, without using oxygen as a source of air enrichment
3 Wear goggles that conform to BS 679

Working on tanks

'Tank' refers to any storage holder, e.g. tank, vessel, drum, container, etc. A very serious danger exists where tanks are to be cut or welded with a flame torch. The contents of the container must be established before any work is considered. All other alternatives should be reviewed before the decision to use flame torches is taken. The risk of explosion is great, and its effect upon the employee or any other person in the vicinity very damaging. The following must be observed for cutting or welding any tank, vessel, drum or container:

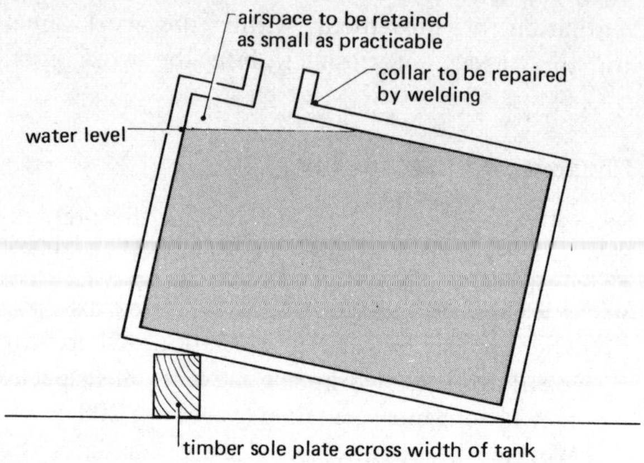

Figure 136 *Precautions when welding a tank*

1 Ascertain what the tank contains or has contained
2 Clean the tank completely of any inflammable or explosive contents by steaming or boiling
3 Never use oxygen to clean the internal part of the tank
4 Fill the tank with water up to a level of 0.050 m from the top. This must also maintain a well-ventilated space at the top part of tank, above the water line (Figure 136)
5 Never use, or allow to be used, any naked flame in the vicinity
6 Erect a warning notice to keep other persons clear of the work area

Working within confined spaces

Dangers that may occur in confined space working are inhaling dangerous fumes, suffocation, or burns. The following must be observed to ensure welding operatives are not endangered:

1 Always keep cylinders outside the confined space. If, for any extreme reason, this is not possible, all cylinders within the confined space must be removed at the end of each working day. This avoids any build-up of gases should a leak occur

2 Always wear a life line when working alone and arrange a system of clear signals with the person assisting with it
3 Maintain a good supply of air for adequate ventilation
4 Never use oxygen to improve ventilation, as this increases the hazard of fire
5 Always have a knowledgeable assistant to control the cylinders from outside and maintain a good back-up service
6 Do not work in a confined space without a fire-fighting aid being easily available – extinguisher or water. See pages 160-5 for details

Working above normal ground level

Where work requires staging or a temporary platform, additional hazards exist to both welder and other personnel. The following must be observed:

1 Always secure the cylinders against falling or bumping
2 Always take precautions to prevent sparks or molten metal from falling on to other persons
3 If protection cannot be achieved, erect a good barrier with clear warning notices
4 Maintain all safe working procedures applicable to platforms above ground level (pages 102-24)

Electric arc welding

Many of the hazards of oxy-acetylene welding apply to electric arc welding.

Materials used, their storage and handling

Electric arc welding uses electrical power to operate the electrode and a return source to the power supply for regeneration. The essence of this scheme is the use of electrical apparatus, and materials are generally less hazardous than with oxy-acetylene procedures.

Cables and couplings

Here is a checklist to ensure safe working conditions:

1 All cables and couplings must be of adequate construction for the welding operations, e.g. rough surfaces, tough conditions
2 All terminals and live components must be sufficiently protected
3 Do not use damaged or poorly insulated cables
4 Avoid unnecessarily long cables
5 Do not allow cables to trail across walkways or similar hazardous work areas
6 All joints must be completed with suitable insulated cable couplings

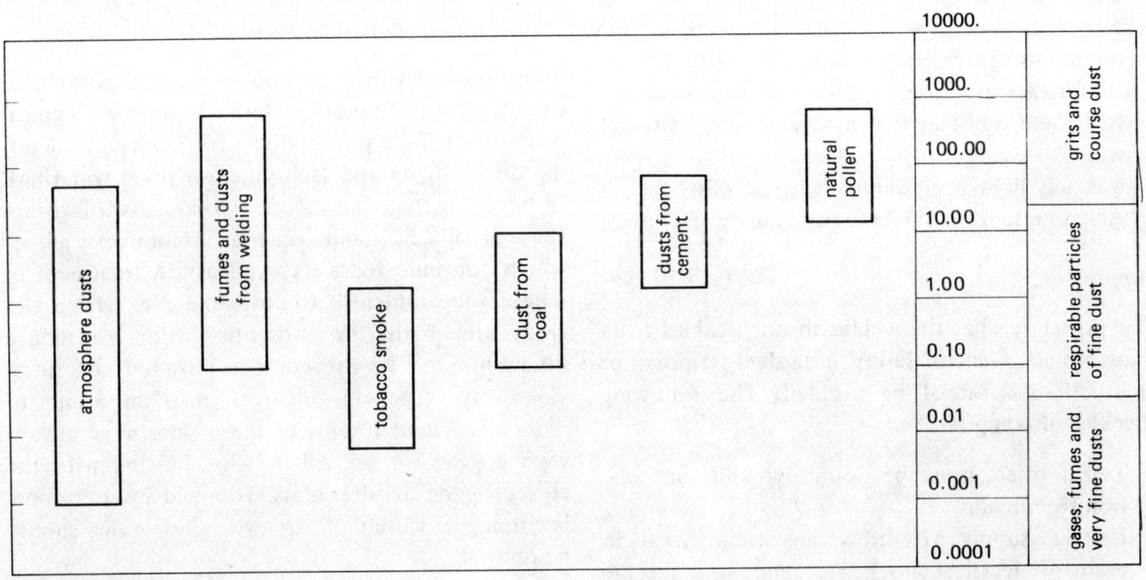

Figure 137 *Particle sizes*

Earthing clamps

Connecting the welding return and welding earth is of primary importance. These prevent the current returning to the source by the wrong circuit, a serious danger. The following must always be observed:

1 Never attempt to connect bare strands of cable to a workplace by clamping to a bolt head or similar fixing
2 Do not pull on any clamping device
3 Always use proper cable lugs and/or earthing clamps

Electrode holders

This apparatus holds electrodes (of any size) whilst the operator works. The following checklist applies to both manufacture and use of electrode holders:

1 Never use a holder that does not have good insulated handles
2 The holder must be balanced when connected to cables and not be too heavy
3 Avoid a rigid cable connection that restricts working
4 Do not leave the holder near, or in contact with, any metalwork when not in use, and always unplug the supply
5 Never pass the holder through apertures, or hang them through holes by the cable with a 'live' circuit. Always turn off the circuit prior to this
6 Never use a holder that has been allowed to get wet
7 Use only holders complying with BS 638
8 Always retain the shield in position on the holder

Supply

The power reaches the welder through cables from transformers. General safety hazards – tripping or obstructions – must be avoided. The following checklist also applies:

1 Transformers must be double wound for isolation from mains
2 Use d.c. supply. (It throws any victim away in event of electrical shock: a.c. tends to make the victim grip the fault and continue the danger)
3 Never use direct mains voltage
4 Only use equipment installed in accordance with IEE *Wiring Regulations* and relevant British Standards.

Preparation of equipment

Provided the equipment selected for use is satisfactory, the following checklist maintains acceptable safety standards:

1 Always ensure a good system of isolation by a fuse switch on any transformer or motor generator
2 Always use interlocking fuse switches for supply to trailing cables to avoid accidental disconnection
3 Never work without a suitable efficient earth connection
4 Check all connections and inspect cables daily

Personal protection

The protection to hands, feet, head and body mentioned for oxy-acetylene welding (page 208) is equally necessary with electric arc welding where molten metal is activated. In addition there are important protective measures for the 'arc' procedure, as eyes are particularly vulnerable.

'Arc-eye' and similar dangers

The ultra-violet light created by electric arc welding causes a painful sensation termed 'arc-eye'. Typical reactions include photophobia (difficulty in tolerating direct light) and lachrymation (eyes watering). In addition there may be temporary loss of vision, acute headaches and general discomfort, all of which continue for a day at least. A treatment to relieve the problem is to cover the eyes with a wet cloth and bathe in antiseptic lotion to reduce inflammation. To prevent the infra-red and ultraviolet rays reaching the eyes, it is important to reduce the light intensity: the welder must always wear goggles, or use a hand shield fitted with the correct grade of filter glass. To avoid any personnel becoming a victim of 'arc-eye', these rules should be obeyed:

1 Never weld without adequate eye protection

2 Isolate the area with non-reflecting (matt green) screens to shield all personnel
3 Avoid any reflective surfaces in the vicinity
4 Never strike an arc without warning those near by who may be unprotected
5 Always erect notices warning of welding flash hazard, 'Warning: arc welding in progress'
6 Never use the equipment in any way that may endanger yourself or any passer-by

Radiation

The same ultra-violet and heat rays created from welding cause skin problems comparable with those from over-exposure to direct sunlight. To avoid this the following should be observed:

1 Always keep bare skin protected from direct rays
2 Provide screens to protect passers-by
3 Display suitable warning notices

Work operation/activity, hazards and precautions

The following operations, restrictions and problems applicable to oxy-acetylene welding apply to electric arc work also: clear working space, ventilation, combustible material hazards (pages 207-11).

26 First aid

The application of first aid covers a vast range of medical topics and is worthy of its own publication for construction employees. This chapter conveys the very basic principles that can be applied, should an accident or misadventure occur. The function of first aid is to care for a casualty until recovered or placed under medical care. The priorities of first aid are:

1 To sustain life by resuscitation or controlling bleeding, etc.
2 To avert the deterioration of any condition by covering wounds, etc.
3 To promote a good recovery by protecting and treating further as necessary.

The first aider

The responsibility of any first aider is to:

1 Assess the problem and situation
2 Diagnose the illness or injury from knowledge gained at the incident or from personal documents' identification
3 Administer immediate and suitable treatment
4 Arrange for further medical treatment as necessary

The requirements of the first aider are to:

1 Be confident and direct, not to hesitate
2 Be reassuring and comforting
3 Be quick in emergency situations
4 Instruct others as necessary – to call a doctor, etc.

Bleeding and cuts

Where an injury has been sustained the first aider must act quickly in the following ways:

1 Get pressure on to the wound by the application of a dressing and pressing directly upon the dressing with finger tips, for up to fifteen minutes
2 Raise the injured area and support it, unless an adjacent fracture is suspected
3 Clear off any dirt or foreign bodies that are easily visible and easily removed
4 Apply a padded dressing and secure it firmly with bandages, ensuring the complete coverage of the wound

Where direct pressure does not control the bleeding, indirect application can be successful as follows:

1 Locate the nearest pressure point between the wound and the heart. This is any location at which an artery can be compressed against an adjacent bone
2 Press on to this artery with a pad, pressure bandage, or directly by hand, to reduce blood loss at the wound
3 This pressure is required until a dressing can be applied, but must not continue for longer than fifiteen minutes

Where serious injuries occur to the chest or abdominal areas, first aid treatment consists of a simple protection against excessive blood loss and infection problems. A clean dressing should be applied and the casualty watched continually until medical aid arrives. Where minor or superficial cuts occur they may be:

1 A clean or incised cut caused by sharp tools
2 Laceration caused by jagged machine parts, metals, or wire
3 Punctured wound caused by small pointed instrument into the body. These situations, if treated incorrectly or ignored, can develop

into a serious problem. Generally, if treated properly, as suggested below, they heal with no more problem

1 Apply a suitable dressing to prevent any further injury, eliminate infection, and control bleeding
2 All dressings must be clean and porous, and first aid administered with clean hands to avoid germs entering the wound
3 Allowance for evaporation of perspiration must be made to reduce moisture build-up and subsequent bacterial infection
4 Adhesive dressings must be applied to dry skin, must allow for perspiration and must be perfectly clean. They should not pull or tear the injury when removed
5 Non-adhesive dressings, consisting of a sterile pad and roller bandage should completely cover the wound They should retain a tight covering to keep out infection but allow the injured area to breathe
6 When bandaging, first apply a clean pad over the wound, then bandage approximately two-thirds of this layer. Pressure should be maintained and the rolled bandage pulled tight during bandaging, but still allowing for body movement beneath
7 Temporary or improvised bandaging can be made with clean and germ-free linen or towelling cut into strips

Fractures

A fracture is a bone cracked or broken. The symptoms are:

1 Swelling and tenderness to the injured area
2 Pain around the injury, regardless of movement
3 Deformed structure of the body
4 Loss of movement caused by a break in body function

If unsure whether a fracture has occurred, always treat as if there is a fracture. The following applies:

1 Avoid movement of the casualty
2 Retain him/her in position and keep warm until medical aid arrives
3 If the casualty *must* be moved, the injured parts must first be strapped to adjacent body parts with a minimum of movement and padding as necessary

Burns

Burns are caused by: dry heat; electricity; friction from wire or revolving apparatus; chemicals, acids or alkalis. Treatment is as follows:

1 Reduce the heat around the burn by careful dabbing with cold water, or immersion, if practicable, for several minutes
2 Make allowance for swelling by removing belts, shoes and jewellery
3 Prevent infection by removing wet clothing but do not remove dry, burned, clothes
4 Cover the burn with large, clean dressing
5 Provide small quantities of cold drink and reassure as necessary
6 Never add risk of infection by touching or contaminating the burns
7 Do not apply oil-based lotions or dressing
8 If the burn is of a chemical nature, remove all contaminated clothing and flush the area clean with cold water before applying a dry dressing
9 Never apply an adhesive dressing to any type of burn

Eye obstructions

Whenever a foreign particle gets under the eyelids, there exists a potentially dangerous situation. The following must be observed:

1 Do not attempt to remove the obstruction from the pupil of the eye, or if it is seriously embedded
2 Stop the casualty from rubbing the affected eye
3 Close the eye and cover with a clean pad until medical aid is available or obtained

Resuscitation

The basic requirements of resuscitation is to provide oxygen into the lungs. The oxygen then circulates in the blood to the brain. If the body is starved of oxygen for more than four minutes, the brain ceases to function, suffers severe damage, and death may occur. If for any reason the casualty stops breathing one of the following methods of resuscitation must be implemented:

Mouth to mouth resuscitation

The simplest form of replacing the oxygen into the lungs, although a high content of carbon dioxide is also given by the rescuer. This imbalance of carbon dioxide to oxygen is a disadvantage but much else can be said in favour of this technique. Mouth to mouth resuscitation is achieved as follows:

1 Tilt the casualty's head backwards, and the chin upwards, to ensure a clear passage for air
2 If this has no effect, there is serious need to apply resuscitation. Take a deep breath, then hold the casualty's nostrils and inflate his lungs by blowing into his mouth
3 Relax from this and at the same time gently press down on the casualty's chest to enforce exhaling
4 Repeat the inflations at normal breathing rates (Figure 138).

Automatic resuscitation

As the name suggests, an automatic resuscitator provides oxygen to the casualty's lungs and then aids the exhaling actions. The process is as follows: cess is as follows:

1 Place the resuscitator eompletely over the mouth and nose
2 Tilt the head backwards and the chin upwards to ensure a good airway
3 If vomiting occurs, turn the casualty on to his side and, after vomiting has ceased, clear the mouth and replace the resuscitator
4 Turn on the appliance and aid the casualty to inhale and exhale as required for recovery

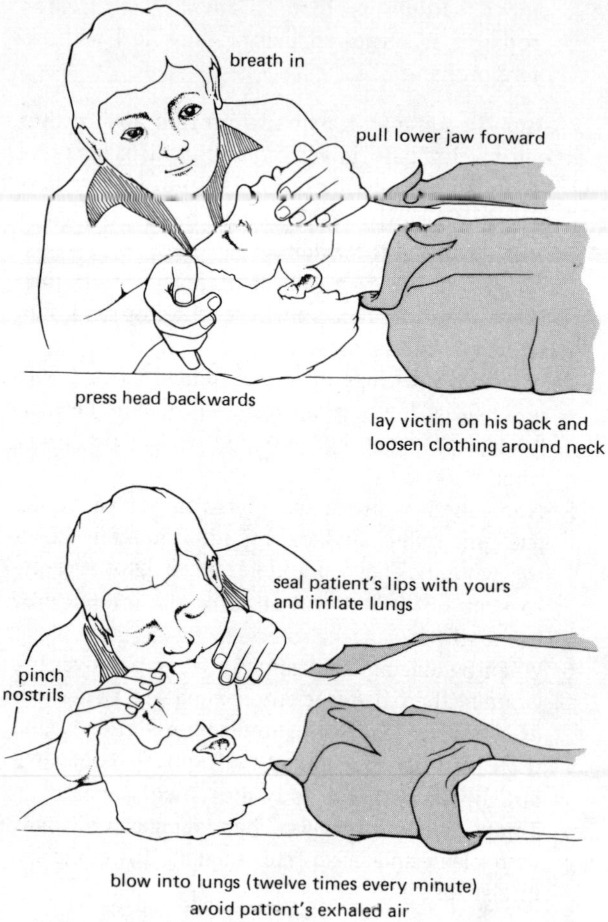

Figure 138 *Mouth to mouth resuscitation procedure*

Guidance Note from the Health and Safety Executive

Metrication of construction safety regulations

General series / 2 (July 1976)

These Guidance Notes replace the series of Technical Data Notes produced by HM Factory Inspectorate and the CEMA's Notes of Guidance produced by the Employment Medical Advisory Service.

The Notes are published under five subject headings; Medical, Environmental Hygiene, Chemical Safety, Plant and Machinery and General.

Existing TDN and CEMA's Notes of Guidance will be progressively brought into the new groupings.

INTRODUCTION

It is likely that some time will elapse before the Construction Regulations made under the Factories Act can be amended. The National Federation of Building Trades Employers, the Federation of Civil Engineering Contractors and the Oil and Chemical Plant Constructors' Association therefore asked the Department of Employment to issue simple advice to contractors who are working on sites where metric linear and weight measurements are in use. For such contractors exact metric equivalents to the imperial measurements in the Regulations may be unworkable, as many run to several decimal places. For example, 6 feet 6 inches = 1·9812 metres.

This note is designed to meet this request and is purely advisory in character. The table suggests rounded-up and rounded-down metric figures which contractors who are using metric measurements may find useful. The principle adopted for linear measurements is generally to round off to the nearest 10 millimetres, but to show measurements of 6 inches or less to the nearest 1 millimetre. All of the figures are within the minimum requirements of the present Construction Regulations. Anyone working to them is therefore complying with the law. One inevitable result of this, however, is that in a few cases a different metric equivalent has been produced for the same imperial figure in different parts of the table, e.g. 6 feet 6 inches in Regulations 9(1) and 27(3) of the Construction (General Provisions) Regulations 1961 has been shown with metric equivalents of 1·980 metres and 1·990 metres respectively. This is because in one case the wording of the Regulation reads 'more than six feet six inches' and in the other case 'at least . . . six feet six inches'.

In order to avoid confusion, the words 'metre' and 'millimetre' are used in full in this note. There is no objection to the verbatim reproduction of the tables in this note provided an acknowledgement of source and Crown copyright is made.

FURTHER INFORMATION

This Guidance Note is produced by HM Factory Inspectorate. Further advice on this and other publications produced by the Executive is obtainable from any Area Office of the Health and Safety Executive, or from HMSO book shops.

TABLE 1

The Construction (General Provisions) Regulations, 1961
Statutory Instrument 1961 No. 1580

Regulation No.	Imperial Requirement	Metric Equivalent (to comply)
8(1)(a)	4 feet	1·210 metres
9(1)	6 feet 6 inches	1·980 metres
9(4)(a)	4 feet	1·210 metres
13	6 feet 6 inches	1·980 metres
27(1)	20 yards	18·280 metres
27(3)	3 feet 6 inches	1·070 metres
	6 feet 6 inches	1·990 metres
28	6 feet 6 inches	1·980 metres
	3 feet	920 millimetres

TABLE 2

The Construction (Lifting Operations) Regulations 1961 Statutory Instrument 1961 No. 1581	Regulation No.	Imperial Requirement	Metric Equivalent (to comply)
	12	2 feet	610 millimetres
	13(1)(c)	6 feet 6 inches	1·980 metres
		3 feet	920 millimetres
		8 inches	210 millimetres
	13(2)	27 inches	680 millimetres
	14(4)(c)	1 ton	1 tonne
	17	6 feet 6 inches	1·980 metres
	28(1)	1 ton	1 tonne
	28(2)	1 ton	1 tonne
	29(1)(b)	1 ton	1 tonne
	30(4)(c)	1 ton	1 tonne
	41(a)	½ inch	13 millimetres
	42(1)	6 feet 6 inches	1·990 metres
		3 feet	920 millimetres
	47(3)(c)(ii)	3 feet	920 millimetres

TABLE 3

The Construction (Lifting Operations) Regulations 1961 Certificate of Exemption No. 1 Hoists in Certain Chimneys (FORM 2006)	Paragraph	Imperial Measurement	Metric Equivalent (to comply)
	Application	120 square feet	11·160 metres^2
	First Schedule Condition 2	80 feet per minute	400 millimetres per second
	Second Schedule Condition 1(a)	3 feet	920 millimetres
		27 inches	680 millimetres
	Second Schedule Condition 3(a)	8 feet	2·440 metres
		2 feet	610 millimetres
	Second Schedule Condition 3(b)	7 feet	2·140 metres
	Second Schedule Condition 4(b)	2 feet 6 inches	770 millimetres
	4(d)	7 feet	2·140 metres
	6(a)	6 feet	1·830 metres

TABLE 4

The Construction (Working Places) Regulations 1966
Statutory Instrument 1966 No. 94

Regulation No.	Imperial Measurement	Metric Equivalent (to comply)
13(5)	3 feet 3 inches	990 millimetres
	1¼ inches	32 millimetres
	5 feet	1·520 metres
	1½ inches	40 millimetres
	8 feet 6 inches	2·590 metres
	2 inches	51 millimetres
15(4)	2 feet	600 millimetres
19(13)(b)(i)	25 inches	640 millimetres
19(13)(b)(ii)	34 inches	870 millimetres
19(13)(b)(iii)	12 inches	300 millimetres
19(14)	10 feet 6 inches	3·200 metres
19(15)	17 inches	440 millimetres
20(1)(g)	3 feet	920 millimetres
21(2)(a)	15 feet	4·570 metres
22(1)	6 feet 6 inches	1·980 metres
24(1)	6 feet 6 inches	1·980 metres
24(1)(a)	6 square inches	3870 millimetres^2
24(1)(b)	1 inch	25 millimetres
24(3)(b)	4 inches	100 millimetres
25(1)(b)	8 inches	210 millimetres
	2 inches	51 millimetres
	6 inches	155 millimetres
25(5)	24 inches	610 millimetres
26(1)	6 feet 6 inches	1·980 metres
26(1)(a)	25 inches	640 metres
26(1)(b)	34 inches	870 millimetres
	17 inches	440 millimetres
26(1)(c)	25 inches	640 millimetres
26(1)(d)	42 inches	1·070 metres
26(1)(e)	51 inches	1·300 metres
26(1)(f)	59 inches	1·500 metres
26(2)	17 inches	440 millimetres
26(2)(a)	25 inches	640 millimetres
	or 34 inches	870 millimetres
26(3)	12 inches	300 millimetres
27(1)	6 feet 6 inches	1·980 metres

TABLE 4 (contd)

Regulation No.	Imperial Measurement	Metric Equivalent (to comply)
27(1)(a)	17 inches	440 millimetres
27(1)(b)	25 inches	640 millimetres
28(1)	6 feet 6 inches	1·980 metres
	3 feet	920 millimetres
	3 feet 9 inches	1·150 metres
	6 inches	155 millimetres
28(3)	30 inches	760 millimetres
28(5)(a)	27 inches	690 millimetres
28(6)(i)	34 inches	870 millimetres
28(6)(f)	6 feet 6 inches	1·980 metres
29(2)	6 feet 6 inches	1·980 metres
29(2)(a)	3 feet	920 millimetres
	3 feet 9 inches	1·150 metres
29(2)(b)	6 inches	155 millimetres
	30 inches	760 millimetres
32(5)(a)(i)	3 feet 6 inches	1·070 metres
32(8)	30 feet	9·140 metres
	6 feet 6 inches	1·980 metres
	3 feet	920 millimetres
	3 feet 9 inches	1·150 metres
	6 inches	155 millimetres
	30 inches	760 millimetres
33(2)	6 feet 6 inches	1·980 metres
33(2)(a)	3 feet	920 millimetres
	3 feet 9 inches	1·150 metres
	6 inches	155 millimetres
	30 inches	760 millimetres
33(4)	6 feet 6 inches	1·980 metres
35(4)(b)	17 inches	440 millimetres
35(7)	6 feet 6 inches	1·980 metres
36(1)	6 feet 6 inches	1·980 metres

TABLE 5

	Paragraph	Imperial Requirement	Metric Equivalent (to comply)
The Construction (Working Places) Regulations, 1966 Certificate of Exemption No. 8 (F 2485) Steeplejacks etc. scaffolds	Second Schedule Paragraph 9	17 inches	440 millimetres

HMSO 30p net ISBN 0 11 883041 4

30 m 8/76

Index